99ECO1-2T

PHASE II

SILVER BIOTIC LIGAND MODEL (BLM): REFINEMENT OF AN ACUTE BLM FOR SILVER

by:

Paul Paquin
Dominic M. Di Toro
HydroQual Inc.

2008

WERF

The Water Environment Research Foundation, a not-for-profit organization, funds and manages water quality research for its subscribers through a diverse public-private partnership between municipal utilities, corporations, academia, industry, and the federal government. WERF subscribers include municipal and regional water and wastewater utilities, industrial corporations, environmental engineering firms, and others that share a commitment to cost-effective water quality solutions. WERF is dedicated to advancing science and technology addressing water quality issues as they impact water resources, the atmosphere, the lands, and quality of life.

For more information, contact:
Water Environment Research Foundation
635 Slaters Lane, Suite 300
Alexandria, VA 22314-1177
Tel: (703) 684-2470
Fax: (703) 299-0742
www.werf.org
werf@werf.org

This report was co-published by the following organizations. For non-subscriber sales information, contact:

IWA Publishing
Alliance House, 12 Caxton Street
London SW1H 0QS, United Kingdom
Tel: +44 (0) 20 7654 5500
Fax: +44 (0) 20 7654 5555
www.iwapublishing.com
publications@iwap.co.uk

Library of Congress Catalog Card Number: 2007936062
IWAP ISBN: 978-1-84339-796-0

HydroQual Incorporated, ENSR, EcoTox, University of Miami, FL, McMaster University, Ontario

ACKNOWLEDGMENTS

The project team extends our thanks to WERF for making these investigations possible and to the International Imaging Industry Association (I3A) and Eastman Kodak Company for the financial support they provided. The technical guidance provided by the Project Subcommittee (PSC) over the course of the project is also gratefully acknowledged. A special thank you is extended to PSC member Joseph Gorsuch, formerly of Eastman Kodak Company and now with Gorsuch Environmental Management Services, Inc. (GEMS), for providing guidance and information from the Silver Research Program.

Report Preparation

Principal Investigators
Dominic M. Di Toro
Paul R. Paquin
HydroQual Inc.

Project Team
Kuen Bing Wu
Rooni Mathew
Aaron Redman
Robert C. Santore
HydroQual Inc.

Rami Naddy
ENSR

Kevin Brix
EcoTox

Martin Grosell
Gretchen Bielmyer
University of Miami, FL

Russel Bell
McMaster University, Hamilton, Ontario, Canada

Scott Smith
Wilfred Laurier University, Ontario, Canada

Project Subcommittee

Norman E. LeBlanc, Chair
Hampton Roads Sanitation District

William J. Adams, Ph.D.
Rio Tinto

Charles Delos
Russell Erickson, Ph.D.
U.S. EPA

Joseph W. Gorsuch
Eastman Kodak Company
Gorsuch Environmental Management Services, Inc. (GEMS)

James R. Kramer
McMaster, University, Hamilton ON, Canada

E. Michael Perdue. Ph.D.
Georgia Institute of Technology

Water Environment Research Foundation Staff

Director of Research: Daniel M. Woltering, Ph.D.

ABSTRACT AND BENEFITS

Abstract:

The biotic ligand model (BLM) is a computational tool that is used to predict toxic effect levels of metals, including silver (Ag). The Ag BLM considers the effect of site-specific water quality characteristics on Ag speciation and bioavailability. This provides a basis for extrapolating from laboratory waters, used for most Ag toxicity tests, to natural waters, where the bioavailability and toxicity of Ag may differ. The original version of the BLM for the acute toxicity of Ag was reviewed by the U.S. EPA Science Advisory Board (SAB) in 1999 and they recommended the completion of additional Ag chemistry and toxicity studies to provide data to test and refine the model. This project was completed to address the SAB recommendations. A chemistry component was carried out to improve the representation of Ag speciation in natural waters and to characterize the important effect of sulfide on Ag bioavailability. These investigations added to the information that was available to characterize Ag chemistry and provided a basis for updating the earlier version of the Ag BLM.

The BLM was also calibrated to laboratory toxicity data that were generated for the Ag-sensitive invertebrate, *Ceriodaphnia dubia*, a cladoceran that had previously been tested to only a limited degree. The studies were carried out with laboratory water in which the chemistry was systematically varied to provide a basis for calibration of the BLM to the toxicity data for *C. dubia*. Toxicity tests were also performed with both *Pimephales promelas* and *C. dubia* in natural water samples that were well characterized chemically, with the data then used to test the model. While the BLM was relatively successful at simulating the observed effect levels for *C. dubia*, the *P. promelas* results were more difficult to interpret and called for further investigation.

Physiological tests were performed to explore the apparent increase in sensitivity of *P. promelas* in some of the test waters that were relatively low in ionic strength. It was determined that acclimation to low ionic strength water, in the absence of Ag, did not significantly reduce their sensitivity to Ag. Rather, it appeared that any benefit that may have been realized by acclimation, such as an enhanced ability to up-regulate ion uptake rates in low ionic strength waters, was apparently offset by a concomitant increase in the rate at which Ag was accumulated. The Ag BLM was modified to incorporate an empirical adjustment as a way to account for the effect of low ionic strength water on the sensitivity to Ag of > 4-day old *P. promelas*. This led to an improved ability of the BLM to predict *P. promelas* effect levels in the natural waters.

Physiological tests were also performed to investigate the apparent species-specific protection against Ag toxicity that is provided by chloride. Investigations with *P. promelas* suggested that the differences in degree of protection may be related to differences in how Ag affects the ionoregulation of Na^+ and Cl^- by different fish species. Although questions remain with regard to chloride, it appears that the BLM is able to predict this response reasonably well, particularly for the relatively sensitive <4 day old *P. promelas*.

In summary, the WERF program yielded an improved chemistry and toxicity database, including data over an extended range of water quality characteristics and for an additional Ag-sensitive invertebrate. These data provided a basis for refining the acute Ag BLM and improving its overall predictive ability.

Benefits:

- The Ag BLM investigations lead to an improved understanding of the bioavailability and toxicity of Ag to aquatic organisms.
- Demonstrates that the ability of the Ag BLM to predict site-specific effect levels in aquatic environments has been improved.
- Provides an expanded database for use in derivation of a BLM-based Ag WQC.
- The U.S. EPA has indicated its intent to incorporate the Ag BLM into an updated WQC for Ag.
- Documents results of the WERF Ag BLM project in seven manuscripts that will be available to WERF subscribers. (See Appendix A for a list of manuscripts.)

Keywords: Silver, biotic ligand model, bioavailability, *Ceriodaphnia dubia, Pimephales promelas*

TABLE OF CONTENTS

Acknowledgments iii
Abstract and Benefits v
List of Tables ix
List of Figures x
List of Acronyms xi
Executive Summary ES-1

1.0 Introduction 1-1

2.0 Silver Chemistry 2-1
- 2.1 Introduction 2-1
- 2.2 Inorganic Ag Speciation 2-2
  - 2.2.1 Silver-Sulfide Chemistry 2-3
- 2.3 Complexation with NOM 2-5
  - 2.3.1 Silver Speciation Measurements 2-5
  - 2.3.2 Calibration to Weak NOM Binding Sites 2-6
- 2.4 Summary 2-7

3.0 Toxicity Tests in Modified Lab Waters 3-1
- 3.1 Introduction 3-1
  - 3.1.1 BLM Calibration Approach 3-1
  - 3.1.2 BLM Version 3-3
- 3.2 *Ceriodaphnia dubia* Test Results 3-3
- 3.3 *P. promelas* Test Results 3-6
- 3.4 Analysis of the Proof of Principle Tests Results and Protection by Sulfide 3-10
  - 3.4.1 Proof of Principle Experiment with *D. magna* 3-11
  - 3.4.2 Proof of Principle Experiment with Rainbow Trout 3-13
- 3.5 Summary 3-14

4.0 Toxicity Tests in Natural Waters 4-1
- 4.1 Introduction 4-1
- 4.2 Natural Water Testing and Related Investigations 4-1
  - 4.2.1 Dataset 1: *C. dubia* in Natural Waters and Water with Reconstituted NOM 4-1
  - 4.2.2 Dataset 2- *C. dubia* and *P. promelas* in Natural Waters from North America 4-5
- 4.3 Summary 4-14

5.0 Discussion, Conclusions and Recommendations 5-1
- 5.1 Introduction 5-1
- 5.2 Discussion 5-1
  - 5.2.1 Interactions of Ag with Sulfide and NOM-binding Sites 5-2
  - 5.2.2 Effect of Size/Age on Sensitivity to Silver 5-5
  - 5.2.3 Effect of Ionic Strength on Sensitivity to Silver 5-6
  - 5.2.4 Effect of Cl^- on Silver Toxicity 5-7
  - 5.2.5 Overall Consistency and Applicability of the BLM 5-12

5.2.6 Are Increasing Concentrations of Competing Cations and Complexing Ligands Protective Against Ag Toxicity? ... 5-13
5.3 Progress Towards the Achievement of Management Objectives ... 5-14
5.4 Conclusions and Recommendations ... 5-15

Appendix A: List of Project Manuscripts ... A-1

Appendix B: Laboratory and Natural Water Test Results – Data Summary ... B-1

LIST OF TABLES

2-1 Inorganic complexation reactions used in the Ag BLM (ionic charges are omitted for clarity) 2-2

2-2 Inorganic Sulfide Reactions Included in the BLM 2-4

2-3 Metal-Sulfide Reactions Included in the BLM 2-5

3-1 Summary of Biotic Ligand Binding Constants for Model with NOM Strong Binding Site Density = 2 nmol/mg DOC 3-2

3-2 Summary of Biotic Ligand Lethal Accumulation Levels for Model with NOM Strong Binding Site Density = 2 nmol/mg DOC 3-2

5-1 Comparison of BLM-predicted and Measured Ag LC50 Responses in Tests with Systematically Varied Chloride 5-10

LIST OF FIGURES

ES-1 Schematic of the Ag BLM Framework ES-2

ES-2 Comparison of Measured (filled and unfilled diamonds) and BLM-predicted (solid and dashed lines) Responses of *C. dubia* LC50 to Na^+ and Cl^- ES-5

ES-3 Relationship of *P. promelas* Na^+ Influx Rate with Weight and Ag Sensitivity with Age or Weight (inset) ES-6

ES-4 Predicted versus Measured Natural Water Test Results for *C. dubia* (left) and *P. promelas* (right) ES-7

1-1 Schematic of the Ag Biotic Ligand Model Framework 1-2

2-1 Effect of Chloride on Ag Speciation (Total Ag = 1 µg/L, pH = 7.0, H = 85 mg/L as $CaCO_3$, Alkalinity = 65 mg/L as $CaCO_3$, DOC = 0.0001 mg/L, Na^+ = 26 mg/L, BL = biotic ligand) 2-3

2-2 Ag titrations of DOM 2-7

3-1 Comparison of Measured (unfilled and filled plot symbols, ± 95% CL) and BLM-predicted (dashed and solid lines) Responses of *C. dubia* LC50 to Na^+ and Cl^- 3-5

3-2 Comparison of Measured (unfilled bars) and BLM-Predicted (filled bars, without any effect of IS on LA50) Responses of 4-day old *P. promelas* LC50 to Cl^- (added as KCL), DOC (added as AHA) and Hardness (added as $CaSO_4$ and $MgSO_4$, on a 1:1 mass basis) 3-7

3-3 Variation of *P. promelas* Na^+ Uptake Rate with Size and Sensitivity to Ag with Age 3-10

3-4 Analysis of Proof of Principle Experiment 3-12

4-1 Predicted versus Measured Natural Water (Dataset 1) and Lab Water LC50 Results for *C. dubia* and *D. magna* (Inset) 4-3

4-2 Preliminary BLM-Predicted versus Measured LC50 Values for *C. dubia* (left) and *P. promelas* (right) for Natural Water Dataset 2, without Metal Competition 4-7

4-3 Comparison of BLM-Predicted and Measured LC50 Values for *C. dubia*, Natural Water Dataset 2 4-8

4-4 Comparison of BLM-Predicted and Measured LC50 Values for *C. dubia*, Composite Dataset 4-9

4-5 Comparison of BLM-Predicted and Measured LC50 Values for *P. promelas*, Natural Water Dataset 2, with 2 nmol/mg DOC but without an IS Adjustment of the LA50 4-10

4-6 Effect of Acclimation Water Quality on 3-hour Gill Ag Uptake in Soft Water by *P. promelas* 4-12

4-7 Predicted LC50 versus Measured LC50 for *P. promelas* Exposed to Ag in Natural Waters and in Laboratory Waters (with IS adjustment except 4-day old fish) ... 4-13

LIST OF ACRONYMS

AHA	Aldrich Humic Acid
BL	Biotic Ligand
BLM	Biotic Ligand Model
CRS	Chromium Reducible Sulfide
DOC	Dissolved Organic Carbon
EC50	Effects concentration for 50% of organisms
IS	Ionic Strength
LA50	Lethal accumulation of a metal at the site of action at 50% mortality
LC50	Lethal concentration for 50% of organisms
MDR	Mixed Diamine Reagent
MT	Metallothionein
NKA	Na^+,K^+-activated ATPase
NOM	Natural Organic Matter
POC	Particulate Organic Carbon
POTW	Publicly Owned Treatment Works
RE	Rotary Evaporation
RO	Reverse osmosis
SAB	Science Advisory Board
TOC	Total Organic Carbon
U.S. EPA	United States Environmental Protection Agency
WER	Water Effect Ratio
WERF	Water Environment Research Foundation
WWTP	Wastewater Treatment Plant
WQC	Water Quality Criterion

EXECUTIVE SUMMARY

Elevated concentrations of metals that result from discharges to aquatic environments have the potential to cause adverse effects on biota. The U.S. EPA has developed methods for deriving ambient water quality criteria (WQC), allowable concentrations that are intended to be protective of aquatic life, in an effort to protect these resources (Stephan et al., 1985; the "Guidelines"). Many of the early WQC for metals were expressed as functions of water hardness in recognition of the perceived effect of increasing hardness towards mitigating the toxicity of some metals (e.g., silver and copper). While this approach was particularly useful because it reflected site-specific conditions (i.e., the hardness of the site water), it was limited in its ability to quantify the influence of other water quality characteristics that were increasingly being recognized as also being of importance. Specifically, steady advances were being made in the understanding of how site-specific water quality characteristics (e.g., pH, alkalinity and DOC) affect the form of a metal that is present (i.e., the metal species) and how these different metal species will in turn affect aquatic organisms. It is now well understood that it is the concentration of these individual metal species, rather than the total metal concentration, that determines the potential for a metal to cause adverse effects. The free ion activity of the metal (e.g., Ag^{+} for silver or Cu^{2+} for copper; the effective concentration after accounting for the concentrations of other ions in the water) was the early focus of attention in this regard. This was because it was found to be the metal species that best correlated to the degree of the toxic effect that was elicited by the metal. This conceptual view is now commonly referred to as the Free Ion Activity Model of metal toxicity, the FIAM (Morel, 1983; Campbell, 1995).

The free ion activity of a metal is often only a small fraction of the total dissolved metal concentration in natural waters. Its magnitude is controlled by the dissolved metal concentration and the chemistry of the bulk solution. That is, other constituents in the water (e.g., dissolved organic matter or sulfide; "ligands" in the jargon of chemists) may form complexes with the metal (e.g., AgHS), thereby reducing the free ion activity and thus its availability to interact with or be taken up by an organism. The situation is further complicated by the observation that other positively charged ions (i.e., cations) that are present in solution (e.g., the hardness cations Ca^{2+} and Mg^{2+}, and also Na^{+} and H^{+}) have a tendency to compete with the positively charged metal ion, thereby further reducing its availability to interact at the site of toxic action. This is why the WQC for some metals were expressed in terms of hardness. In summary then, and as succinctly described by Playle (1998), "the three Cs" that need to be considered in assessing metal bioavailability are concentration (of the reactive species), complexation and competition.

At the time that the FIAM was first formally described (Morel, 1983) and for some time thereafter it was not obvious how to incorporate the FIAM concepts into the regulatory process. Pagenkopf (1983) used a mathematical model to simulate measured toxicity levels for several metals, but the approach was never adopted for use, perhaps because it failed to provide a tangible demonstration of a key underlying principle, the importance of quantifying the level of accumulation at the site of toxic action. Playle and co-workers highlighted the path forward when they used a chemical equilibrium model to simulate the preceding effects of water chemistry on the experimentally determined metal accumulation levels (initially for Cu, Cd and Ag) at the proximate site of action of acute metal toxicity to fish, the gill (Playle et al., 1992, 1993a,b; Janes and Playle, 1995). MacRae and co-workers subsequently provided evidence that the level of gill metal accumulation was directly related to the degree of the toxicological response of an organism to a metal (MacRae et al., 1999). This served as a concrete demonstration of the somewhat simplistic but useful approximation that a fixed level of metal accumulation at the site of toxic action of toxicity, the biotic ligand, is associated with a fixed effect. Di Toro et al. (2001) made use of the preceding results to predict site-specific dissolved metal effect levels in the context of a chemical equilibrium model, given information about the chemistry of the water of interest (Figure ES-1, as applied to Ag). The model was based upon CHemical Equilibria in Soils and Solutions (CHESS), an inorganic speciation model (Santore and Driscoll, 1995), modified to incorporate the representation of metal-organic matter

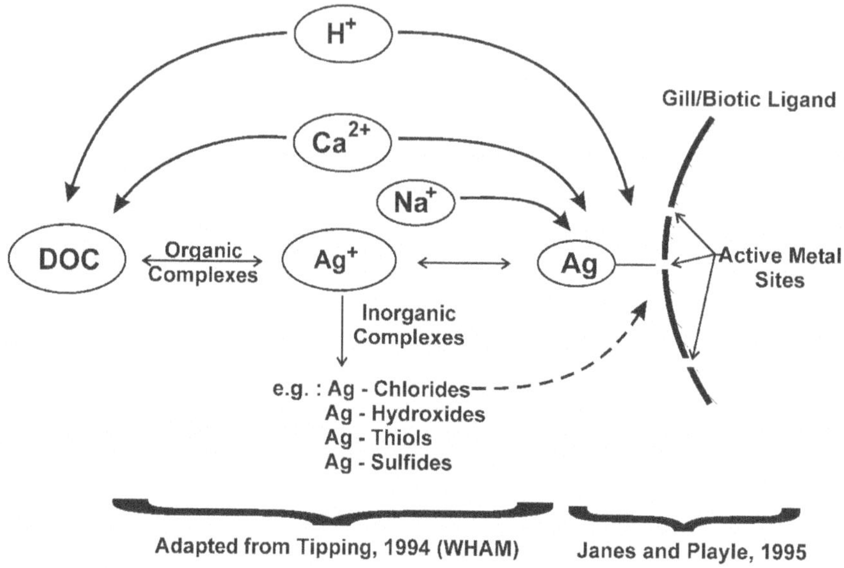

Figure ES-1. Schematic of the Ag BLM Framework (Wu et al., 2008).

interactions used in the Windermere Humic Aqueous Model (WHAM; Tipping, 1994). This so-called biotic ligand model (BLM) was well suited for use in the regulatory arena because it was designed to predict toxic effect levels for metals (e.g., the 96-hour LC50, the dissolved metal concentration that is lethal to 50% of the test organisms in a 96-hour period), the information that serves as the basis for how WQC are commonly derived.

The initial efforts to introduce the BLM into the regulatory process included the presentation of early versions of the Cu and Ag BLM to the U.S. EPA Science Advisory Board (SAB) (U.S. EPA, 1999a). Pursuant to their review (U.S. EPA, 2000) the SAB stated: "... *independent verification of the model in predicting acute silver toxicity with additional datasets is recommended to identify other ligands (e.g., sulfide) that form complexes with silver, and to assess the accuracy of the model under a broader spectrum of water quality conditions and aquatic organisms. We encourage the regulatory and scientific communities as they further apply the BLM model to collect paired datasets of water-effect ratio measurements and BLM measurements for copper, silver, and other metals to allow for model improvement and validation*". The WERF Ag BLM chemistry, physiology and toxicology studies and related modeling analyses have helped to satisfy these needs for Ag (a similar set of investigations were performed for Cu). An overview of the study results is provided below, with more detailed descriptions provided in the body of this report and in a series of project-related manuscripts (Bielmyer et al., 2007, 2008; Naddy et al., 2008; Santore et al., 2008; Smith et al., 2002, 2004; Wu et al., 2008).

ES.1 Summary of Chemistry Results

As stated above, the WHAM framework provided the basic structure for how interactions between metals and natural organic matter (NOM), quantified as dissolved organic carbon (DOC), were represented in the BLM. Although WHAM had previously been calibrated for a wide variety of metals and types of NOM, it had not previously been applied to silver. An important component of the WERF project was therefore to chemically characterize the interaction of Ag with natural water samples, including samples of concentrated NOM (Smith et al., 2002 and 2004). These test results were then used to refine the representation of Ag-organic matter interactions that had previously been incorporated into the BLM on the basis of limited information (Santore et al., 2008). The BLM was also used to analyze toxicity datasets where elevated levels of NOM and sulfide (an inorganic ligand that forms a strong complex with Ag) were present in the test waters. This provided a further test of the model and the way that the complexation reactions of Ag with these constituents were represented (Santore et al., 2008). A concentration of strong binding sites of 2 nmol/mg DOC was incorporated in the BLM on the basis of the analysis of Ag toxicity data, including the analysis of the natural water toxicity test results described below. This refinement improved the overall LC50 response of the model to DOC, particularly for the relatively sensitive invertebrate species.

The results of the preceding analyses, when used in combination with the stability constants for inorganic Ag speciation that are available in the NIST database, provide a relatively complete representation of Ag speciation in natural waters over a range of typically encountered water quality characteristics: pH, alkalinity, Cl^- and DOC. It turns out that alkalinity is a relatively unimportant factor with respect to silver speciation. Chloride is important at environmentally relevant concentrations, because increasing the chloride concentration from 1 to 300 mg/L (~0.03-8.50 mM) can reduce the free Ag from more than 90% to less than 5% of the dissolved Ag when the DOC concentration is low (< 1 mg/L). As discussed below, because AgCl appears to be bioavailable to aquatic organisms, its formation does not entirely mitigate adverse effects due to Ag, though some protection does appear to be provided by an increasing Cl^- level. At a relatively high Cl^- level of about 100 mg/L or greater, $AgCl_2^-$ begins to form. Because this complex is thought to be unavailable to the organism, the formation of this chloro-complex of Ag provides a less ambiguous level of protection against toxicity due to Ag than does AgCl. The pH of the water has a limited effect on Ag speciation in low DOC waters (< 1 mg/L), because Ag does not tend to form hydroxy- or carbonato-complexes. It does increase in importance as the DOC increases, however, because the decrease in proton (i.e., H^+) concentration with increasing pH reduces the binding of H^+ ions to NOM, effectively increasing the complexation capacity of the NOM for Ag^+. It is for this reason that DOC is expected to have a somewhat limited effect on silver speciation over a pH range of 5-6 in comparison to its effect over a pH range of 8 - 9.

Interestingly, calcium, which has historically been considered to have a protective effect against metal toxicity, is relatively ineffective at mitigating the toxicity of Ag. Further, it may even increase Ag toxicity in high DOC water, because Ca^{2+} can reduce the complexation capacity of NOM by competing with Ag^+ for NOM binding sites. Sulfide, which is normally present at low concentrations in natural waters, forms a strong complex with Ag^+, thereby reducing the concentration of the bioavailable Ag by what appears to be a 1:1 molar basis. The high propensity of sulfide to bind Ag is important, given that the environmentally low levels of sulfide that are present in most natural waters may still be sufficient to complex the typically low levels of Ag that might otherwise be toxic to some Ag-sensitive aquatic organisms. To the degree the sulfhydryl groups are associated with NOM, the effect of NOM on silver speciation is enhanced.

ES.2 Summary of Laboratory Toxicity Test Results

Use of the BLM to evaluate silver bioavailability and effect levels involves more than the evaluation of silver speciation because the interaction of silver with the biotic ligand also needs to be considered. The biotic ligand is represented by specifying its nominal concentration (i.e., the binding site density) in the water in the same way that the concentrations of other complexing ligands are specified. Biotic ligand binding constants for the metal of interest and for other competing cations must also be assigned, and these are typically obtained by calibration of the BLM to both accumulation data (if available) and to toxicity data. For the WERF analyses described herein, the BLM had previously been calibrated to the gill Ag accumulation results of Janes and Playle (1995), following a procedure first outlined by them (see Wu et al., 2008 for the evaluation of the biotic ligand stability constants). However, because the primary use of the Ag BLM is to predict Ag effect levels rather than Ag accumulation levels the binding constants were subsequently refined to achieve a reasonable simulation of the available Ag toxicity data. The biotic ligand-Ag accumulation level that is associated with 50% lethality, the LA50, is also evaluated from the toxicity data. The Ag^+ binding constants are considered to be independent of organism type (consistent with the underlying cause of Ag toxicity being the inhibition of the same enzyme systems across organism types). Conversely, the LA50 is assumed to vary by species as a way to represent differences in species (and/or age class) sensitivity. The Ag LC50 for a water sample that has a prescribed set of water quality characteristics is predicted by simply evaluating the dissolved metal concentration in the water (the LC50) that will result in a Ag-biotic ligand (Ag:BL) concentration that is equal to the LA50.

The WERF laboratory toxicity investigations focused on tests with *C. dubia*, a silver-sensitive aquatic organism that had not formerly been widely tested in exposures to Ag. The tests were performed in lab water that was modified to achieve different ranges of pH, alkalinity, hardness (Ca^{2+} and Mg^{2+}), Na^+, Cl^- and DOC. These *C. dubia* tests added to the variety of test organisms and widened the concentration ranges of water quality characteristics to which the BLM had previously been applied. The *C. dubia* LA50 was evaluated on the basis of these test results in modified lab waters. An example comparison of toxicity test results with simulated BLM results in tests where Na^+ and/or Cl^- were varied is shown on Figure ES-2. The Ag LC50 is plotted as a function of Na^+. In the first sequence of experiments the Na^+ was varied by the addition of either Na_2SO_4 or $NaHCO_3$. The data (unfilled symbols) indicate a steady increase in LC50 as the Na^+ is increased from 2.5 mg/L to 160 mg/L (~0.10-7.0 mM). The model (dashed line) yields a more gradual response with the predicted increase in LC50 being somewhat less than that which is exhibited by this first set of results. The response is in very good agreement with the LC50 for the lower Na^+ treatment ($Na^+ \sim 40$ mg/L) in the second series of experiments, where the water quality characteristics, including Cl^-, were comparable to that of the first set of tests. The results for the second series of tests, which made use of NaCl to adjust the Na^+ level, also shown on Figure ES-2 (filled diamonds). A somewhat greater increase in LC50 was in this case observed at the highest Na^+ level, presumably due to the covariation of Cl^- with Na^+ in the test water. The BLM results (solid line) exhibit a more gradual response than the data to the increase in NaCl. However, in either case (model or data), the enhanced increase in LC50 in the presence of NaCl is clearly evident.

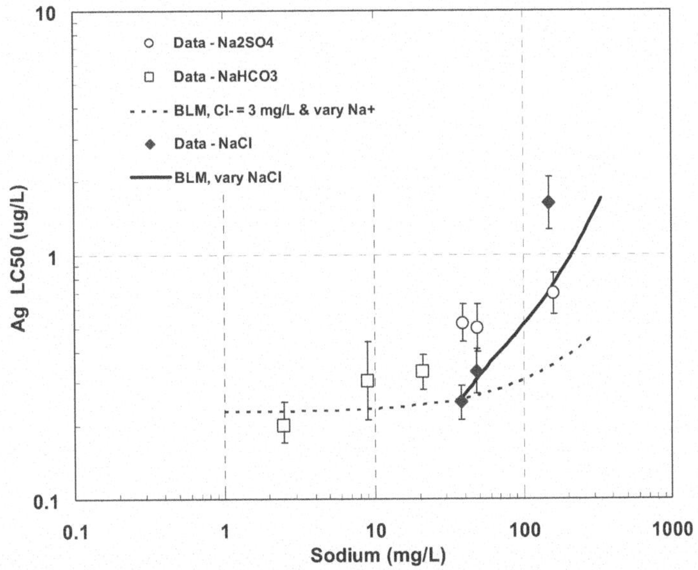

Figure ES-2. Comparison of Measured (filled and unfilled plot symbols, ±95% CL) and BLM-predicted (solid and dashed lines) Responses of *C. dubia* LC50 to Na^+ and Cl^- (Data: Naddy et al., 2008)

Although tests with *P. promelas* (fathead minnow) were not performed in modified lab waters as part of the WERF investigations, *P. promelas* was used to test the ability of the model to predict effect levels for this fish species over a range of natural water test conditions. It is for this reason that calibration results for *P. promelas*, based on previously published test results, are reviewed in Chapter 3.0. One aspect of these analyses that was explored in detail was the effect of size on the sensitivity of *P. promelas* to silver. Consistent with previous results, it was found that the sensitivity to Ag tended to be inversely related to size. That is, the Ag LC50 tends to increase (i.e., sensitivity decreases) as fish age and/or size increases (Figure ES-3, inset). It is known that silver is toxic to aquatic organisms by inhibiting their ability to take up and maintain the required elevated concentrations of Na^+ and Cl^- in their internal fluids. If sufficiently severe, this can have lethal consequences. It has been shown that the variation in species sensitivity to Ag is related to the rate at which internal Na^+ pools are replaced by newly accumulated Na^+, such that a high turnover rate is associated with a high sensitivity to silver (Bianchini et al., 2002b; Grosell et al., 2002). The reason is that if the uptake of Na^+ is reduced by a fixed amount by exposure to Ag^+, an organism with a high turnover rate will experience a net loss of Na^+ at a faster rate than an organism with a low turnover rate and, as a result, it will be relatively sensitive to Ag^+. It has

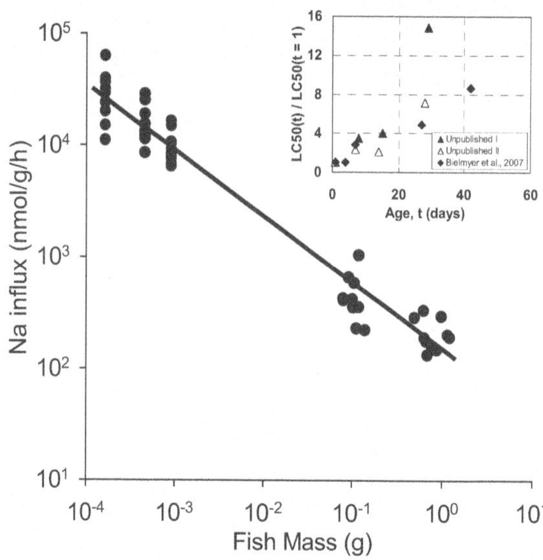

Figure ES-3. Relationship of *P. promelas* Na⁺ Influx Rate with Weight and Ag Sensitivity with Age or Weight (inset) (Data: Bielmyer et al., 2007 and unpublished results)

also been shown that Na^+ turnover rate tends to be inversely related to organism size (Bianchini et al., 2002b; Grosell et al., 2002). Sodium influx measurements that were made with different size *P. promelas* as part of the WERF investigations showed that this same sort of relationship between Na^+ uptake rate and size also applies to *P. promelas* sensitivity and size. It follows, from a BLM perspective, that size-related differences in the sensitivity of *P. promelas* can be represented in the same way as differences in species sensitivity, by adjusting the LA50.

The question of whether or not Cl^- provides a uniform level of protection against Ag toxicity for all species was also addressed as part of the WERF investigations. The reason for doing so was that some datasets indicated that increasing the chloride level by 1 mM or more provided some level of protection (e.g., Figure ES-2) while other tests failed to demonstrate a significant increase in LC50 with increasing Cl^- (Bielmyer et al., 2008). Physiological tests highlighted the possibility that species-specific differences occur as a result of the differential effects of Ag on Cl^- uptake (Bielmyer et al., 2008). The discrepancy may also have reflected the statistical power of the test design and the difficulty of detecting a relatively small change in LC50 with a standard test design, one that includes only three replicates. It is also possible that the potential for an increase in LC50 with increasing Cl^- was offset by the increase in K^+ that occurred in many of the datasets considered (i.e., in tests where KCl was used to adjust the Cl^- level), given the limited evidence showing that K^+ can adversely effect Na^+ regulation (e.g., Wilcox and Dietz, 1995). Regardless of the answers to these questions, the BLM, as currently calibrated, tends to slightly understate the protective response to Cl^- that is evident in most of the datasets where Cl^- levels were systematically varied, including the response of the relatively sensitive 4-day old *P. promelas* to Cl^-.

ES.3 Summary of Natural Water Test Results

The SAB called for the independent verification of the Ag BLM with test data that represented a "broader spectrum of water quality conditions" relative to the data available as of 1999 (U.S. EPA, 2000). Toxicity tests were therefore performed in a variety of natural waters.

Dataset 1 included toxicity tests that were performed with *C. dubia* exposed to Ag in natural water samples, wastewater effluent samples, and lab water that was amended with reconstituted NOM concentrates that were obtained from these same site waters (by rotary evaporation, RE, or reverse osmosis, RO). Dataset 2 included tests with *C. dubia* and 8-day old *P. promelas*, with each exposed to Ag in natural water samples from eight different locations in the United States and Canada.

The BLM calibration for *C. dubia* is mainly based upon test results in modified lab waters that were low in NOM (Naddy et al., 2008). The current version of the model also includes NOM strong binding sites at a concentration of 2 nmol/mg DOC, based on the analysis of tests with both NOM-amended lab waters and the natural water results of Dataset 2 (discussed in further detail in Chapter 4.0). The BLM-predicted versus measured *C. dubia* results are shown on the left panel of Figure ES-4. The graph includes both the modified lab water results (unfilled diamonds), Dataset 1 results for lab waters adjusted to site water hardness levels (filled diamonds) and the natural water results for Dataset 2 (i.e., site water results, the filled triangles). The middle diagonal line is the line of perfect agreement between model and data while the outer diagonal lines define the region within which the predictions are within a factor of two of the data.

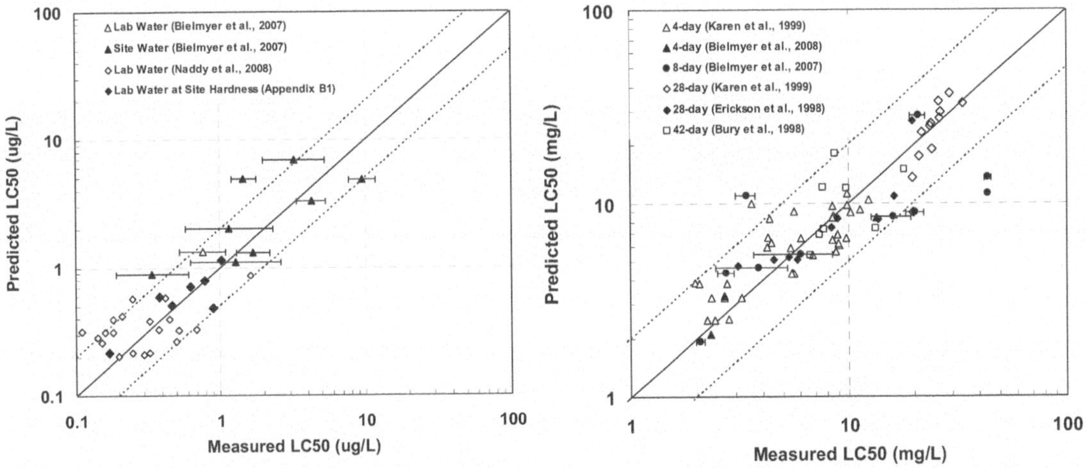

Figure ES-4. Predicted versus Measured Natural Water Test Results for *C. dubia* (left) and *P. promelas* (right)

The water samples that were tested in association with Dataset 1 had relatively high levels of chromium reducible sulfide (CRS), organic matter and chloride in comparison to the natural water samples of Dataset 2. However, the Ag LC50 values for Dataset 1 were considerably less than would be expected on the basis of the CRS complexation capacity that was estimated from stoichiometric considerations. Thus the results that were predicted by the BLM were consistently a factor of two or more above the data (see Figure 4-1). It was concluded that the problem was related to the limited equilibration time that was provided between the test water and the added silver prior to the addition of the test organisms (generally less than 1 hour, versus 24 hours for Dataset 2). This conclusion is supported by recently published toxicity test results in the presence of elevated levels of NOM (added as Aldrich humic acid, AHA) which showed that an equilibration time that was longer than one hour was probably required for the complexation reaction to approach equilibrium (Glover et al., 2005a). It was inferred from the responses of *D. magna* to Ag following 3 and 24 hour pre-equilibration times (Glover et al., 2005a) that the rate constant for the complexation reaction is only about 20 to 30% per hour. While the complexation reactions would be expected to continue during the test itself, recent results have highlighted the important influence of water chemistry during the initial portion of a toxicity test on the resulting effect levels that are measured (Naddy et al., 2007; Kolts et al., 2006). These findings are qualitatively consistent with the results of BLM simulations that used only 20% of the measured CRS and 20% of the assigned NOM strong binding site concentrations to predict LC50 values for Dataset 1 that were comparable in magnitude to the data.

In spite of the complications associated with the interpretation of the results for Dataset 1, these data provide a clear and unambiguously conservative demonstration of the potential for natural water characteristics to mitigate the toxicity of Ag to aquatic organisms. Water effect ratios (WER = LC50 in site water / LC50 in lab water adjusted to site water hardness) calculated with these data ranged from 3.0 to 11.6. The values would likely have been higher had the equilibration time been longer. The BLM, which had previously been calibrated to low NOM lab water toxicity data for *C. dubia* (unfilled diamonds on left panel of Figure ES-4), was used to estimate how much higher the WERs might have been had a longer equilibration time been used in the tests. Based on these results it was estimated that the WERs would have ranged from 19.3 to 53.5 under equilibrium conditions. While WER values of this magnitude may seem quite high, they are entirely consistent with the WER of 60.3 that was measured with *D. magna* (lab water LC50 = 0.58 µg/L), an organism of comparable sensitivity to *C. dubia* (Dataset 1 average LC50 = 0.58 µg/L and range of 0.21 - 1.03 µg/L for the hardness adjusted lab waters). This WER was measured in St. Louis River water that had been pre-equilibrated with Ag for 72 hours prior to the addition of the test organisms (Erickson et al., 1998). The high value of this WER reflected the high DOC of the site water (18.3 mg/L) and perhaps an elevated CRS (CRS was not actually measured). While the reference water had not been adjusted to the hardness of the St. Louis River water, this factor was not likely important given the slight effect of hardness on Ag toxicity to *D. magna* (Chapter 3.0).

The predicted and measured *C. dubia* results for Dataset 2 are also compared on the left panel of Figure ES-4 (filled triangles, with 95% confidence limits). As mentioned above, the initial Dataset 2 results for *C. dubia* (Bielmyer et al., 2007) were also improved by refinements that were made to the BLM calibration over the course of the WERF investigations, including the incorporation of 2 nmol/mg DOC of strong binding sites in combination with an adjustment to the LA50. As shown on Figure ES-4 the predictions of *C. dubia* LC50 values in the natural water samples are in relatively good agreement with the measured LC50 values. Most results for the overall composite dataset for *C. dubia* are within a factor of two of the measured LC50 values.

Application of the same chemistry representation that was used for *C. dubia* yielded initial BLM results for the 8-day old *P. promelas* (Dataset 2) that responded less than the data to changes in the water quality characteristics of the test waters (Bielmyer et al., 2007 and Chapter 4.0). The difficulty seemed to be related to the low ionic strength of some of the water samples (IS < 0.01 M). A similar response to the ionic strength had previously been observed in lab water test results with *P. promelas*, except for the youngest age class that had been tested, 4-day old fish (Wu et al., 2008). Acclimation of the 8-day old fish to the low IS test water for 7 days prior to their exposure to Ag did not lead to a significant change in the measured LC50 results. This was thought to be due to the demonstrated increase in gill silver uptake rate by the fish that were provided time to acclimate to the low ionic strength test water (Bielmyer et al., 2008).

The BLM-predicted response for the 8-day old *P. promelas* was improved by incorporating an empirical adjustment to the LA50 in low IS water. While a similar adjustment was also called for in the BLM analyses of supplemental toxicity datasets for older age *P. promelas*, for reasons that are not well understood it did not seem to be needed for 4-day old *P. promelas*, the most sensitive age class tested. Results obtained with the modified LA50 inputs are compared to the data for 8-day old fish, exposed to silver in the 8 natural water samples, on the right panel of Figure ES-4 (filled circles). Related *P. promelas* results for other age classes (mostly from earlier investigations in modified lab waters; except the filled triangles) are also shown on this graph. As shown, the natural water test results span about a 20-fold range in variation, from about 2 to 40 µg/L. In contrast to the initial BLM results the model in this case simulates the observed LC50 values reasonably well for most of the natural water samples, including the lower effect levels. It does tend to underestimate two of the observed LC50 values by considerably more than a factor of two when the LC50 is high, on the order of 40 µg/L. While there are a few exceptions, the LC50 values for the other *P. promelas* age classes are predicted reasonably well, with the predictions generally within a factor of two of the data.

ES.4 Results Related to the EPA SAB Recommendations and the Achievement of Management Objectives

Pursuant to their 1999 review of the Ag and Cu BLM, the U.S. EPA Science Advisory Board identified the following needs related to the continued development of the BLM: "*... independent verification of the model in predicting acute silver toxicity with additional datasets is recommended to identify other ligands (e.g., sulfide) that form complexes with silver, and to assess the accuracy of the model under a broader spectrum of water quality conditions and aquatic organisms. We encourage the regulatory and scientific communities as they further apply the BLM model to collect paired datasets of water-effect ratio measurements and BLM measurements for copper, silver, and other metals to allow for model improvement and validation.*", (U.S. EPA, 2000). The WERF Ag BLM chemistry, physiology and toxicology studies and related modeling analyses have helped to satisfy these needs. The Ag chemistry studies provided detailed test results that have improved our understanding of silver speciation in natural waters, particularly with regard to interactions with natural organic matter and with inorganic sulfide, water quality variables that the WERF studies and other investigations have shown to be important factors that affect the bioavailability and toxicity of silver to aquatic life. The improved representation of Ag chemistry was used, in combination with paired toxicity test results (i.e., tests performed with natural water samples and laboratory waters) and physiological investigations, to develop an improved capability to predict acute Ag effect levels over broader concentration ranges of the water quality constituents that affect silver bioavailability and effects than was previously available for use. These tests and modeling analyses were performed with *P. promelas*, a previously well-tested fish, and with *C. dubia*, a cladoceran that had previously received limited attention but one that is very sensitive to Ag at relatively low environmental exposure levels (Figure ES-4).

The WERF Ag BLM project represented one aspect of a multi-faceted effort, an effort that had the support of the regulatory, industrial and scientific communities, to gain an improved understanding of the bioavailability and effects of metals, including Ag, in the environment. It was important in this regard that the results of the WERF Ag investigations would support the achievement of a number of important management objectives. These objectives included the following: (i) the attainment of an improved understanding of the effects of site-specific water quality characteristics on metal bioavailability and effects, (ii) the development of models that were not overly complex that would satisfy the needs of both the regulatory (e.g., permit writers) and regulated end-users (i.e., dischargers) and (iii) the establishment of scientifically sound acute and chronic WQC for freshwater and saltwater environments.

Considerable progress has been made toward the achievement of these management objectives as a result of the WERF Ag, Cu and Ni BLM efforts, in combination with ongoing EPA and industry sponsored initiatives (primarily for Ag, Al, Cd, Cu, Ni, Pb and Zn). These efforts have led to important advances in the current understanding of how to assess the effect of water chemistry on the bioavailability and toxicity of Ag and other metals. The data from many of these studies has also provided a basis for the development of a user friendly version of the BLM, with implementations currently available for Ag, Al, Cd, Cu, Ni, Pb and Zn. EPA has spearheaded efforts to incorporate these results into updated and scientifically defensible site-specific WQC that reflect these advances in understanding and make use of the modeling tools that have been developed. The 2007 BLM-based WQC document for Cu provides the most tangible demonstration of this (U.S. EPA, 2007). Pursuant to the completion of the WERF Ag BLM studies, as well as related investigations that have been completed in the context of the broader Silver Research Program (SRP), EPA has recently initiated parallel WQC-development efforts for Ag and Zn. The objective of these efforts is to develop BLM-based WQC for these metals. As is the case for other metals, further work remains to be completed in regard to advancing the state of the art for the derivation of technically defensible marine and chronic WQC for Ag. However, a greatly enhanced body of information and additional data are now available that will be of direct use in the achievement of these remaining longer-term management objectives (Paquin et al., 2007).

ES.5 Conclusions and Recommendations

The following conclusions are made pursuant to the results presented herein:

- Based on the consistency of the BLM speciation results with the available Ag speciation data, and the ability of the model to reproduce the organism response to Ag that is observed in the majority of the toxicity tests, the Ag BLM provides a reasonable estimate of Ag speciation in natural waters.
- The toxicity database for Ag BLM development purposes has been expanded considerably and now includes additional data for *C. dubia* and *P. promelas*, for an expanded range of water quality characteristics.
- The updated acute Ag BLM should serve as an appropriate tool for incorporation into an updated WQC document for Ag. This BLM-based WQC should be amenable to updating as the current understanding of the bioavailability and effects of Ag (and other metals) on aquatic organisms continues to advance.

Initial efforts to validate the BLM with toxicity test data for natural water samples met with mixed results. The *P. promelas* analyses in particular led to the need to refine the BLM to better account for effect levels for *P. promelas* that are more than 4 days old that are exposed to Ag in low ionic strength water.

- The updated model may be used to predict LC50 values for both *P. promelas* and *C. dubia* over a wide range of water quality characteristics, including elevated levels of CRS.
- The representation of ionic strength effects seems to be important for older fathead minnows, but this response does not seem to be significant for the most sensitive age class tested the 4-day old age class.

Given that the available data are subject to alternative interpretations, the general ability of Cl^- to protect against Ag toxicity across species and how to best represent Cl^- protection in the BLM would benefit from further clarification. One view is that the absence of a definitive (i.e., statistically significant) response to chloride in some of the experimental datasets warrants that Cl^- protection be invoked on a species-specific basis. Alternatively, in view of the general consistency between BLM results and the current body of information use of a generic model of Cl^- protection is reasonable. Additional data will help to resolve this uncertainty. The decision to be made at present is whether to proceed with the BLM in its current form, or to wait until future data become available for consideration. A relatively conservative approach would be to continue to use the Ag BLM, but without an allowance for Cl^- protection. A disadvantage of this approach is that it would necessitate a recalibration of the BLM to the available data and doing so could lead to the unintended consequence that compensatory artifacts are introduced into the BLM calibration. Alternatively, given the tendency of the currently calibrated version of the BLM to predict a protective response to Cl^- that is less than is indicated by most of the datasets analyzed (including the 4-day old *P. promelas* age class), it could continue to be used in its present form with additional refinements being incorporated, as appropriate, as new data become available for consideration.

While the effect of age or size on the sensitivity of *P. promelas* to silver is reasonably well understood (i.e., it is a function of the size related differences in sodium turnover rate) this is not the case for the effect of low IS water on sensitivity to silver. In either case, however, the approach used to account for these differential responses was the same, to simply adjust the LA50. When using the Ag BLM for effects assessments and WQC derivation purposes it will not be necessary to employ these corrections if the evaluation is based on the most sensitive age class, the 4-day old fish. That is, a single LA50 that is independent of ionic strength effects could be used for *P. promelas* if that approach is adopted for use. Use of the most sensitive age class is in keeping with the spirit of the 1985 guidelines for the derivation of WQC (Stephan et al., 1985). Specifically, the Guidelines state: "If the available data indicate that one or more life stages are at least a factor of two more resistant than one or more other life stages of the same species, the data for the more resistant life stages should *not* be used in the calculation of the SMAV...". The SMAV, or Species Mean Acute Value, is the effect level that is used in the evaluation of the WQC.

As described herein, the Ag investigations that were carried out have helped to advance that state of understanding of the bioavailability and toxicity of Ag in aquatic environments. As is normally the situation with this type of research effort, however, there are a few key technical areas that would benefit from further study. More specifically, it is recommended that:

- Given the few studies that have failed to identify a statistically significant protective effect of chloride against Ag toxicity, the uncertainty associated with standard test protocols that have been routinely used in the past, and the current representation of a limited protective benefit in the BLM, toxicity tests incorporating a more robust experimental design should be carried out. The standard test design could be enhanced by including: an increased level of replication, a chemistry component that considers Ag speciation, side-by-side comparisons where Cl^- is added with different cations (e.g., Na^+, K^+, Mg^{2+}, Ca^{2+}), physiological measurements of ionoregulatory end-points (e.g., plasma ion levels and one way ion fluxes), confirmation of the toxicity test results by one or more independent laboratories, and modeling.

- The effect of acclimation conditions and of low ionic strength waters on organism sensitivity to metals is an area that will require further exploration if the operative processes are to be fully understood. Given the recognized need for this type of information across metals, generally, these studies should include tests with both fish and invertebrate species and should be performed not only with Ag, but with other metals as well. While the completion of multi-generation studies would also be informative, they would not necessarily be of direct use in furthering the state of development of an acute BLM for silver or other metals. A comprehensive study design should include the following components: chemistry (i.e., a careful characterization of metal speciation), physiology, toxicity and modeling.

CHAPTER 1.0

INTRODUCTION

The significant effect that water chemistry has on metal bioavailability and toxicity has been repeatedly demonstrated (see Campbell, 1995, for a comprehensive review). As a result, it is now generally recognized by the scientific, regulatory and regulated communities that site-specific water quality needs to be considered when setting discharge requirements for metals. However, while water quality criteria (WQC) for many metals incorporate an adjustment for the hardness of the site water, the effects of other water quality characteristics are generally not considered directly. The studies described herein were initiated to address this limitation. The project included both laboratory testing (chemical, physiological, toxicological studies) and modeling, to provide a technically defensible basis for the evaluation of an updated acute water quality criterion (WQC) for silver (Ag) in aquatic environments, one that explicitly considers site water chemistry. The project focused on the completion of studies that would lead to a refined version of the biotic ligand model (BLM), a modeling framework that the United States Environmental Protection Agency (U.S. EPA) is planning to use in the evaluation of updated WQC for several metals, including Ag (U.S. EPA, 1999b). Very briefly, the BLM framework considers how site-specific water quality characteristics affect the bioavailability and toxicity of metals to aquatic life (Di Toro et al., 2000; 2001). Just as the currently applied hardness correction is intended to adjust for the effect of hardness on the toxicity of a number of metals, the BLM framework is designed to also consider the effects of additional water quality characteristics such as pH, alkalinity, hardness chloride and dissolved organic matter.

Given its important relationship to the overall study it is useful to begin with a description of the BLM framework (Figure 1-1). The BLM is a variation of the free ion activity model (FIAM; Morel, 1983) and the gill surface interaction model (GSIM; Pagenkopf, 1983). The FIAM states that the response of an organism to a metal is directly related to the free ion activity (i.e., the effective concentration after correcting for the ionic strength of the media) of the metal rather than to the dissolved metal concentration itself. Morel (1983), in providing the earliest formal description of the FIAM, also recognized the potential for other cations to influence toxicity, via competition with the free metal ion (and possibly other metal complexes) for binding at the site of action of toxicity. The experimental work and early modeling by Playle and co-workers (Playle et al., 1992, 1993a,b; Janes and Playle, 1995) provided the stimulus that was needed to make use of these ideas within a regulatory context.

The version of the BLM employed herein uses a standard chemical equilibrium computational approach to evaluate the concentrations of organic and inorganic Ag complexes that form in waters having variable water quality characteristics (Santore et al., 1995; Di Toro et al., 2001). It is important that this be done as accurately as possible, because metal complexes are known to vary markedly in their bioavailability and toxicity to the organism. Although chemical equilibrium models have been widely and successfully applied in the evaluation of metal speciation in aquatic settings for most metals, the representation of Ag chemistry with such models, particularly at the relatively low toxic effect levels of Ag, has been less well advanced (e.g., Bell and Kramer, 1999; Adams and Kramer, 1999). Hence, a significant component of the work described here was directed at obtaining an improved understanding and representation of the chemical interactions that control Ag speciation in water.

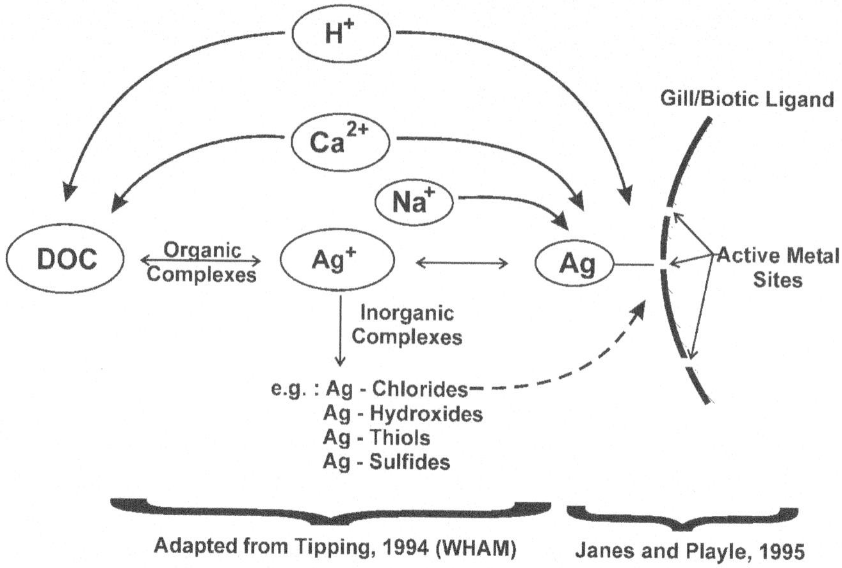

Figure 1-1. Schematic of the Ag Biotic Ligand Model Framework.

With regard to the interaction of Ag with the organism, the BLM represents metal binding at the site of action of toxicity, the gill in the original work with rainbow trout by Janes and Playle (1995), as a complexing ligand (Figure 1-1, right side). The predicted Ag accumulation at this site of action reflects both the distribution of Ag among the chemical species that exist in the water and competition with other dissolved cations for physiologically active binding sites at the gill. Within the context of the BLM framework for acute toxicity (U.S. EPA, 1999a; Di Toro et al., 2001), this calculated level of Ag accumulation is used to predict the dissolved metal concentration associated with a particular response (e.g., the lethal concentration associated with 50% mortality for a fixed exposure duration, the LC50). Because the gill represents the proximate rather than the precise site of action of toxicity (i.e., one or more sensitive intracellular enzyme systems) a more general term, the "biotic ligand", was adopted to describe this site of action. It should also be recognized that water chemistry not only has an effect upon metal speciation, but it may also affect the physiological condition and sensitivity of an aquatic organism to a metal. Such effects are not explicitly represented within the BLM. Rather, they are subsumed within the chemical interactions that are represented in the BLM and the calibration of the BLM itself to toxicity data.

BLM development efforts prior to initiation of the studies described herein had focused on the application of the BLM to the evaluation of acute effect levels for Ag and Cu (Paquin et al., 1999; Di Toro et al., 2000, 2001; McGeer et al., 2000; Santore et al., 2001; Andren and Bober, 2002).[1] The BLM was presented to the EPA Science Advisory Board (SAB) (U.S. EPA,

[1] Efforts to develop versions of the acute BLM for other metals (e.g., Al, Cd, Ni, Pb and Zn) were subsequently initiated (Santore et al., 2002; Brooks et al., 2004; HydroQual Inc., 2001, 2002, 2003; Paquin et al., 2002a), as have applications for assessing chronic toxicity (De Schamphelaere and Janssen, 2004).

1999a), and although the SAB was generally receptive to the approach, a number of areas were identified where further research was needed to refine and further validate the BLM (U.S. EPA, 2000). Quoting from the SAB report: "... *independent verification of the model in predicting acute silver toxicity with additional datasets is recommended to identify other ligands (e.g., sulfide) that form complexes with silver, and to assess the accuracy of the model under a broader spectrum of water quality conditions and aquatic organisms. We encourage the regulatory and scientific communities as they further apply the BLM model to collect paired datasets of water-effect ratio measurements and BLM measurements for copper, silver, and other metals to allow for model improvement and validation.*" The Scope of Work (SOW) for the Ag BLM project incorporated tasks that were designed in part to address these SAB recommendations. The SOW included the following:

- Chemistry studies to characterize the speciation of Ag in natural water samples from waters impacted by effluents from Publicly Owned Treatment Works (POTWs), including the characterization of the binding strength and concentrations of ligands in natural waters, and use of these results to refine the representation of Ag chemistry employed by the BLM (Smith et al., 2002; 2004; Santore et al., 2008;
- Toxicity tests with the cladoceran *Ceriodaphnia dubia*, an invertebrate that is very sensitive to Ag, in lab waters in which the chemistry was systematically varied (Naddy et al., 2008);
- Toxicity tests with the fathead minnow (*Pimephales promelas*) and with *C. dubia* in natural waters that represented a variety of water quality characteristics (Bielmyer et al., 2007; Appendix B, Table B-3) and physiological tests to better understand (i) how the test water characteristics (specifically, acclimation and exposure to Ag in a low ionic strength, IS, soft water) affected *P. promelas* sensitivity and (ii) species-specific differences in the protective effect of chloride (Bielmyer et al., 2008)
- Application of the BLM to toxicity test results from a variety of studies (Bielmyer et al., 2007; Naddy et al., 2008; Wu et al., 2008; Bianchini et al., 2002a; Mann et al., 2004), including (i) toxicity tests with *C. dubia* in laboratory water in which the chemistry was systematically varied, (ii) toxicity tests with *P. promelas* and *C. dubia* in a variety of natural waters; (iii) other Silver Research Program (SRP) test results, including "Proof of Principle" test results for *Oncorhynchus mykiss* (rainbow trout) and *D. magna* (i.e., tests designed to evaluate if sulfide mitigated toxicity due to Ag).

The overall research needs for Ag were quite broad such that not all of the Ag-related studies that needed to be completed were carried out under the auspices of WERF (e.g., Bianchini et al., 2002a and Mann et al., 2004, conducted the "Proof of Principle" tests mentioned above). However, the WERF Ag BLM project was able to build upon the results of many of these parallel investigations, thereby enhancing the overall utility of the results of the WERF investigations (see Paquin et al., 2007, for a description of the overall SRP).

Though not entirely dependent upon the results of the WERF Ag BLM project, it was important for the WERF Ag investigations to provide information that would support the achievement of several important management objectives of both the regulatory and regulated communities. These management objectives included the establishment of scientifically sound acute and chronic WQC for freshwater and saltwater environments, the development of methodologies to account for the effects of site-specific conditions in natural waters, and models that are not overly complex, such that the needs of both the regulatory (e.g., permit writers) and regulated (i.e., dischargers) end-users are satisfied.

This overview report is intended to provide a summary of the WERF project findings. To this end the Executive Summary provides a concise overview of the most important project results, conclusions and recommendations. A more detailed description is provided in the balance of the report, organized in part in accordance with the four main areas of investigation (i.e., chemistry, lab water toxicity tests and natural water testing, with model development and application discussed throughout). Chapter 2.0 focuses on the chemistry investigations, including both experimental investigations aimed at the characterization of Ag speciation in natural waters and the representation of these results in the BLM. Chapter 3.0 describes the toxicity tests in systematically modified lab waters, tests that were performed to broaden the range of water quality characteristics over which the BLM had been applied. These tests were performed with *C. dubia,* an invertebrate that is sensitive to Ag at levels near the proposed draft acute WQC for Ag (0.92 µg/L at hardness = 50 mg/L as $CaCO_3$; U.S. EPA, 1987). Earlier lab water toxicity test results with *P. promelas* are also discussed as a prelude to Chapter 4.0. Chapter 4.0 includes a review of the natural water testing program, designed to obtain data for use in validating the previously calibrated BLM. As will be seen, these tests highlighted some limitations of the original model with regard to its ability to reproduce measured effect levels over the full range of conditions tested. These test results led to the completion of additional physiological investigations into why the deviations between model and data occurred and to modification of the BLM to improve its predictive capability over the range of conditions tested. The discussion of the analysis of the project data with the BLM, the fourth principal area of investigation, is integrated into the discussions of the chemistry, lab water and natural water test results that are reviewed in Chapters 2.0 through 4.0, respectively. Finally, Chapter 5.0 includes a discussion of the project results, focusing on some of the important technical areas that were addressed by the project, and includes conclusions and recommendations that were arrived at on the basis of the results presented herein. Additional information regarding these investigations is included in nine manuscripts that describe various components of this project in considerably more detail than contained herein (as listed in Appendix A: Bielmyer et al., 2007, 2008; Naddy et al., 2008; Santore et al., 2008; Smith et al., 2002, 2004; Wu et al., 2008).

CHAPTER 2.0

SILVER CHEMISTRY

2.1 Introduction

In the BLM conceptual framework, chemical speciation is one of the primary factors that determine metal bioavailability to aquatic organisms (Figure 1-1). Dissolved metals can exist in different chemical forms, depending on their association with one or more ligands. It is assumed that the transformation between these chemical forms is rapid enough that a chemical equilibrium framework based on thermodynamics is a useful description of the distribution of dissolved metal between various complexes. The distribution between different chemical species is dependent on the concentrations of dissolved silver and of ligands capable of forming complexes with Ag. These ligands may be inorganic or organic in nature. Since silver is a type B metal, inorganic complexes are expected to be dominated by interactions with sulfides and chlorides. This distinguishes silver from other metals such as copper, in that while bicarbonate complexes dominate inorganic copper speciation (assumed to be Cu(II) for the purposes of this discussion) in most natural waters (as well as zinc and lead), they are unimportant for silver. Differences are also anticipated between the complexation of silver and copper with organic matter. While copper will bind to carboxylic acids and phenolic functional groups, silver has weak interactions with these groups, yet has moderately strong interactions with amine-based functional groups and very strong binding with sulfur-based functional groups. The conceptual model for the BLM can easily account for these types of differences in the chemical behavior of different metals, and the importance of chemical speciation in determining metal bioavailability is not reduced as a result of these differences.

Although details, such as the predominance of different types of complexes, may differ when comparing the chemical behavior of different metals, the silver BLM has many similarities to the BLM versions for other metals in that chemical speciation is governed by complexation with both inorganic ligands and natural organic matter. A similarity of the silver BLM with versions for other metals is that the broad spectrum of binding sites (weak, moderately strong, to very strong) results in an apparent concentration-dependent binding strength between silver and the available ligands. At low concentrations of silver, the apparent binding may be dominated by the strongest available sites. As silver concentrations increase, these strongest sites may be saturated with silver and unable to participate in further complexation reactions, thereby allowing the next strongest sites to bind silver. The ligands that dominate silver speciation in a given experiment, therefore, depend on both the silver concentration and the availability of ligands (such as may be derived, for example, from natural organic matter). Because the intended use of the BLM is to predict the toxic effects of silver, the concentration range that the model needs to operate at are dictated by the relative sensitivity of the various organisms used in toxicity tests. In order for the model to accurately describe silver speciation in these tests, it needs to include a sufficient diversity of binding sites needed for the spectrum of toxicological responses that include very sensitive organisms (such as *C. dubia*) and relatively insensitive organisms (such as fathead minnow or rainbow trout). While individual chemical experiments may seem to indicate that the speciation of silver can be simplified to a relatively few reactions, it must be clear how the relative silver concentrations in each experiment, as well as that of the important ligands, compares with conditions that are typical in toxicity tests, as well as in real world settings in natural waters where the silver BLM may be used to set regulatory criteria.

2.2 Inorganic Ag Speciation

As has already been mentioned, silver is a type B metal and is therefore expected to form inorganic complexes primarily with sulfides and chlorides. Complexation reactions with bicarbonate and carbonate, which are important for copper, are insignificant for silver. The inorganic speciation reactions included in the silver BLM are based on stoichiometry and thermodynamic information from the NIST database (Martell et al., 2003). Ignoring, for the moment, interactions with sulfide, which will be discussed subsequently, the remaining inorganic speciation reactions are summarized in Table 2-1. Of these reactions, the complexation reactions with chloride are clearly the most important.

Table 2-1. Inorganic Complexation Reactions Used in the Ag BLM (ionic charges are omitted for clarity).

Reaction Stoichiometry					Log K
Ag	+	H_2O	=	Ag(OH)	2.00
Ag	+	$2*H_2O$	=	$Ag(OH)_2$	3.99
Ag	+	Cl	=	Ag(Cl)	3.31
Ag	+	2*Cl	=	$Ag(Cl)_2$	5.25
Ag	+	3*Cl	=	$Ag(Cl)_3$	5.20
Ag	+	4*Cl	=	$Ag(Cl)_4$	5.51
Ag	+	NO3	=	$Ag(NO_3)$	-0.10

Figure 2-1 illustrates the percentages of dissolved silver (1 μg/L ~ 10 nM) that are present as either free silver or as mono- or di-chloro Ag complexes as Cl^- varies (pH = 7 in a low DOC water). At a Cl^- concentration of about 1 - 2 mg/L more than 90% of the Ag is present as Ag^+ with the chloro-complexes of minor importance (~5%). The AgCl fraction increases in importance with increasing Cl^-, accounting for about 60% of the Ag at 35 mg/L Cl^- (1 mM Cl^-) and about 70% at 100 mg/L (~ 3 mM). The Ag^+ and $AgCl_2^-$ fractions each account for about 15% of the silver at this point. Finally, as Cl^- increases to still higher levels, $AgCl_2^-$ becomes increasingly important as the AgCl fraction begins to decrease. Levels of chloride in excess of 100 mg/L are not typically observed in freshwater settings, but could be present in the vicinity of a wastewater discharge or in a stream or river as it transitions to an estuarine region.

Figure 2-1 also illustrates how the change in speciation with varying Cl^- affects accumulation of Ag at the biotic ligand (Total BL:Ag), as it is represented in the Ag BLM (Wu et al., 2008). As shown, the BLM allows both Ag^+ and AgCl to accumulate at the biotic ligand (BL:Ag^+ and BL:AgCl, respectively). As chloride increases and the distribution shifts from Ag^+ to AgCl, total BL:Ag decreases even though both forms of Ag are considered to be bioavailable. The reason is that Ag in the form of Ag^+ will accumulate to a greater degree than when it is present as AgCl. Thus, Cl^- provides a partial protective benefit over this range of Cl^- concentrations, even though Ag in the form of AgCl is available for uptake by the organism. As the Cl^- continues to increase and the distribution shifts to $AgCl_2^-$, a form that is not considered to be bioavailable, an enhanced protective effect is provided by further increases in chloride.

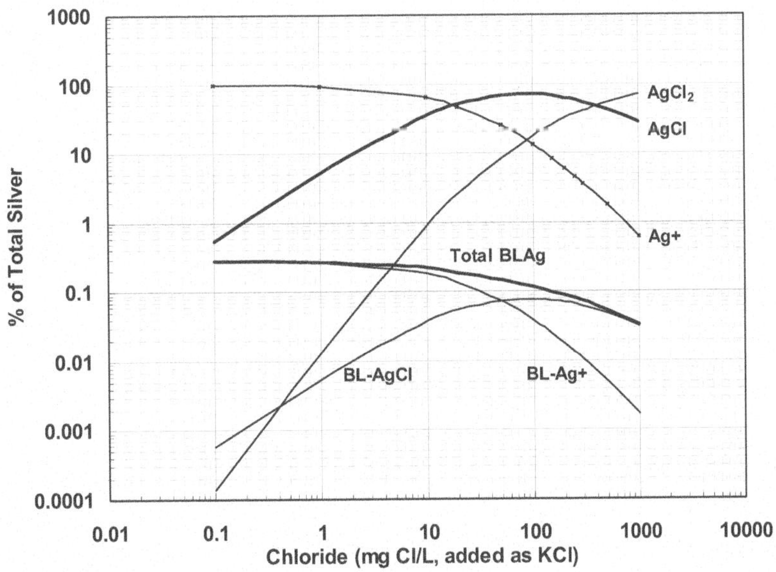

Figure 2-1. Effect of Chloride on Ag Speciation (Total Ag = 1 µg/L, pH = 7.0, H = 85 mg/L as $CaCO_3$, Alkalinity = 65 mg/L as $CaCO_3$, DOC = 0.0001 mg/L, Na^+ = 26 mg/L, BL = biotic ligand)

2.2.1 Silver-Sulfide Chemistry

Sulfide speciation in aquatic systems and its interactions with metals (including silver) have been extensively reported in the literature (see Rickard and Luther, 2006, for a recent comprehensive review of metal-sulfide chemistry). Similarly, Bell and Kramer (1999) have published an analysis and review of silver speciation in the presence of organic and inorganic forms of sulfide. Since both the toxicological exposures in the presence of sulfide, and the analytical method for sulfide measurement employed in the WERF studies involve inorganic forms of sulfide, the subsequent discussions focus on inorganic sulfide chemistry.

Reduced sulfide ligands have been reported in oxic environments at up to sub-micromolar concentrations (Bowles et al., 2002a; Kramer et al., 1999; Rozan et al., 2000; Kuwabara and Luther, 1993). The stability and persistence of sulfide in such oxic environments has been attributed to the formation of complexes with metals such as lead, manganese, cobalt, silver, copper, mercury, and iron (Kramer et al., 2007; Rozan et al., 2000). Inorganic sulfide has also been hypothesized to be associated with natural organic matter (NOM) by either occlusion within the pore spaces of NOM molecules or by binding with functional groups on NOM molecules, thus leading to the development of linear relationships between sulfide and NOM (Kramer et al., 2007).

2.2.1.1 Inorganic Sulfide Speciation

Any analysis or prediction of silver-sulfide speciation involves the determination of the inorganic speciation of sulfide, and the interaction between silver and sulfide. Under typical freshwater conditions, inorganic sulfide speciation follows classical thermodynamic acid-base reactions, forming aqueous $H_2S(aq)$, the hydrosulfide ion HS^- (also called bisulfide) and sulfide ion S^{2-}. The H_2S system is defined by dissociation constants pK_1 = 7.02 (Smith and Martell, 2001; Martell et al., 2003), and pK_2 varying from 13.9 to 18.5 (Morel and Herring, 1993;

Rickard and Luther, 2006), with the uncertainty in pK_2 attributed to the oxidation of sulfides to polysulfides at high pH. In spite of this uncertainty, under typical non-acidic natural environmental conditions, the bisulfide ion can be expected to dominate sulfide speciation. As mentioned previously, sulfide can also exist as polysulfide, formed either as a result of WWTP operations (Rozan et al., 2000), or at elevated pH, or when sulfide is present at millimolar or higher concentrations (Bell and Kramer, 1999). The controlling reactions for inorganic sulfide speciation, as implemented in the BLM, are summarized in Table 2-2.

Table 2-2. Inorganic Sulfide Reactions Included in the BLM.

Reaction	Thermodynamic Constant, Log K	Source
$H^+ + HS^- = H_2S$	7.02	Martell, et al., 2003
$H^+ + S^{2-} = HS^-$	13.9	Morel and Herring, 1993

2.2.1.2 Silver-Sulfide Interactions

Silver can form a number of dissolved and precipitated (solid) species in the presence of sulfide, including colloidal and nano-molecular particles (Bell and Kramer, 1999). The solid forms of silver-sulfide are restricted to acanthite ($Ag_2S(s)$), which is formed at low temperatures (T < 173 °C), and argentite, which tends to form at higher temperatures (Bell and Kramer, 1999).

Information on the dissolved speciation of silver-sulfide systems was determined from the NIST database (Smith and Martell, 2001; Martell et al., 2003) and from a review of recent literature on metal-sulfide interactions. The NIST database of stability constants (Smith and Martell, 2001; Martell et al., 2003), is a compilation of critically selected thermodynamic equilibrium constants from the literature and is a standard reference in the development of chemical speciation models. Therefore, the NIST database was used as the primary source of information on sulfide speciation, with the information obtained from the literature review providing supplemental information. With regard to the literature review, Rickard and Luther (2006) provide a comprehensive review of the literature on metal-sulfide chemistry and have tabulated the various forms of silver-sulfide complexes and their associated thermodynamic binding constants. As described therein, the exact composition of silver-sulfide complexes is the subject of some debate, with some researchers reporting complexes with the bisulfide ion (HS^-) and others reporting complexes with the sulfide ion (S^{2-}). Reported Ag:S stoichiometries for the silver-bisulfide complexes, $(Ag^+)_i(HS^-)_j^{(i-j)}$, include 1:1, 1:2, and 2:3, with the 1:2 stoichiometry complex expected to dominate speciation over pH range 5.8 to 7.3 (Rickard and Luther, 2006). On the other hand, Bell and Kramer (1999) suggest that the 1:1 stoichiometry species dominates silver-sulfide speciation over total sulfide concentrations ranging from $10^{-13.3}$ to $10^{-13.9}$ M and at pH=7. Rickard and Luther (2006) also report silver-sulfide complexes, $(Ag^+)_i(S^{2-})_j^{(i-2j)}$ with stoichiometries of 1:1 and 2:1. However, it is not clear if the various complexes co-exist in a state of thermodynamic equilibrium or if the occurrence of a particular complex is contingent upon the specific experimental conditions such as temperature, pH, or concentrations of silver and/or sulfide.

Silver-bisulfide complexes and thermodynamic stability constants similar to those presented in Rickard and Luther (2006) are also found in the NIST database. Further support for the silver-sulfide reactions presented in the NIST database comes from a number of studies using competitive ligand titrations with silver in the presence of "strong ligands" (considered to be predominantly reduced sulfide ligands) which indicate binding constants and Ag:S stoichiometry similar to the 1:1 stoichiometry AgHS species described in the NIST database (Smith et al., 2004; Smith et al., 2002; Herrin et al., 2001; Kramer et al., 2007). This suggests that the silver-sulfide reactions and constants described in the NIST database may be adequate and appropriate to represent silver speciation in the BLM. The silver-sulfide interactions implemented in the BLM are summarized in Table 2-3.

Silver-sulfide speciation in the presence of other metals has also been studied. Studies conducted in zinc-sulfide cluster systems (Rozan and Luther, 2002) have shown that silver is effective at replacing zinc from the zinc-sulfide clusters, indicating that silver has a strong affinity for sulfide. Similar results were also found with copper-sulfide clusters in the presence of silver. In addition, Luther and Rickard (2005) have calculated the binding constants for silver-sulfide cluster species to be several orders of magnitude greater than for any other metal-sulfide clusters, with the exception of mercury. The ability of silver to out-compete other metals for binding with sulfide is of consequence since most natural water bodies would be expected to contain trace amounts of background metals. Therefore, no interactions between sulfide and any other metal is represented in the BLM (i.e. it is assumed that silver will out-compete any other metal for complexation with sulfide).

Table 2-3. Metal-Sulfide Reactions Included in the BLM.

Reaction	Thermodynamic Constant, Log K (20°C and I = 0.0)	Source
$Ag^+ + HS^- = AgHS$	13.38[2]	Smith and Martell, 2001; Martell et al., 2003
$Ag^+ + 2HS^- = Ag(HS)_2^-$	17.47[2]	Smith and Martell, 2001; Martell et al., 2003
$2Ag^+ + H_2S = Ag_2S_{(s)} + 2H^+$	29.2	Smith and Martell, 2001; Martell et al., 2003

2.3 Complexation with NOM

2.3.1 Silver Speciation Measurements

Natural organic matter is a complex mixture of dissolved and particulate material derived from the decomposition of plants. Dissolved organic matter is typically defined as natural organic matter that will pass through a 0.45 µm filter. This filtration level is coarse enough to include both truly dissolved, and some colloidal and other materials that can be removed with finer filtration sizes. However, for practical reasons 0.45 µm is the most widely used filtration level to define dissolved metals and other substances, and as such will be considered here as the basis for the definition of both dissolved silver, and dissolved organic matter (DOM).

Because of the complex nature of NOM, metal complexation characteristics cannot typically be represented by a single binding site. Successful models of NOM-trace metal complexation typically try to represent a range of binding sites, with corresponding ranges in binding strengths using either a large number of heterogeneous binding sites (e.g., Tipping et al., 1994), or with a probability distribution of binding site behavior (e.g., Kinniburgh et al., 1996). Consideration of a range in binding sites, one that includes relatively stronger and weaker sites, results in an apparent binding strength between NOM and silver that is concentration-dependent. At given NOM concentrations and with low concentrations of silver, the apparent binding may

2 The log K values in the NIST database were reported for elevated ionic strengths. For purposes of this table they were adjusted to be consistent with T = 20°C and I = 0.0.

be dominated by the strongest available sites. As ionic silver concentrations increase, these strongest sites may be saturated with silver and unable to participate in further complexation reactions, thereby allowing the next strongest sites to bind silver. The ligands that dominate silver speciation in a given experiment, therefore, depend on both the silver concentration and the quantity of NOM. This information can be summarized as a silver/NOM ratio. Stronger binding sites would be expected to be more readily apparent at lower silver/NOM ratios, and the apparent binding strength would decrease as the silver/NOM ratio of the experimental conditions increases. Because the BLM will ultimately be used to predict the toxicity of silver, it is important to conduct speciation measurements in conditions that include silver/NOM ratios that are typical in toxicity experiments.

2.3.2 Calibration to Weak NOM Binding Sites

Previous BLM versions have used the Windermere Humic Aqueous Model (WHAM) for simulating complexation between NOM and metals. However, WHAM has not been previously calibrated to include interactions with silver. The binding sites in WHAM include carboxylic and phenolic sites which are expected to have very weak association with silver. Nevertheless, these weak site interactions may be important in toxicity exposures with insensitive organisms, where the quantity of silver needed to elicit a toxicological response overwhelms the available strong binding sites, or in synthetic waters where sources of strong silver binding (such as would be measurable by CRS) are absent. Quantification of weak site interactions requires available silver speciation data in the presence of NOM at very high Ag/NOM ratios.

Although a tremendous amount of work has been published in recent decades on the interactions between NOM and trace metals, very little of this work has been directed at understanding silver speciation. Sikora and Stevenson (1988) measured silver binding with NOM derived from soil extracts, and their results were used as one of the candidate datasets for calibration of weak interactions with NOM. In addition, as part of the WERF project additional data were generated to measure weak interactions between Ag and DOC, using an ion selective electrode (ISE) to measure free Ag ion. Although both of these datasets were considered in the calibration of weak NOM interactions, the data from Sikora and Stevenson are at very high total silver concentrations, and are therefore less relevant to the range of conditions necessary to parameterize the BLM.

Quantification of these weak site interactions is shown for both datasets in Figure 2-2. The weak nature of these sites is evident by the fact that free silver ion concentrations are close to total silver concentrations. Nevertheless, weak site complexation accounts for the small reduction in free silver ion. For calibration of the BLM it is assumed that these weak interactions occur at the carboxylic and phenolic functional groups represented in WHAM. Calibration of the WHAM humic and fulvic interaction results in the model simulations shown in Figure 2-2. Good agreement to the data from Sikora and Stevenson could be obtained with calibrated Ag-proton exchange constants in WHAM for humic and fulvic acids of 2.64 and 2.2, respectively (not shown). However, these calibrated values underestimated the amount of binding observed in the data from Smith and Bell. Calibration to Smith and Bell produced somewhat stronger binding as represented by Ag-proton exchange constants for humic and fulvic acids of 2.4 and 1.7. These values fit the Smith and Bell data well, but tended to overestimate binding in Sikora and Stevenson (Figure 2-2). Using parameters values that represent a good fit to the data from Smith and Bell is preferable because the very high total silver concentrations in the Sikora and

Stevenson experiments (upwards of 10,000 μg/L) make those data less relevant to parameterize the BLM.

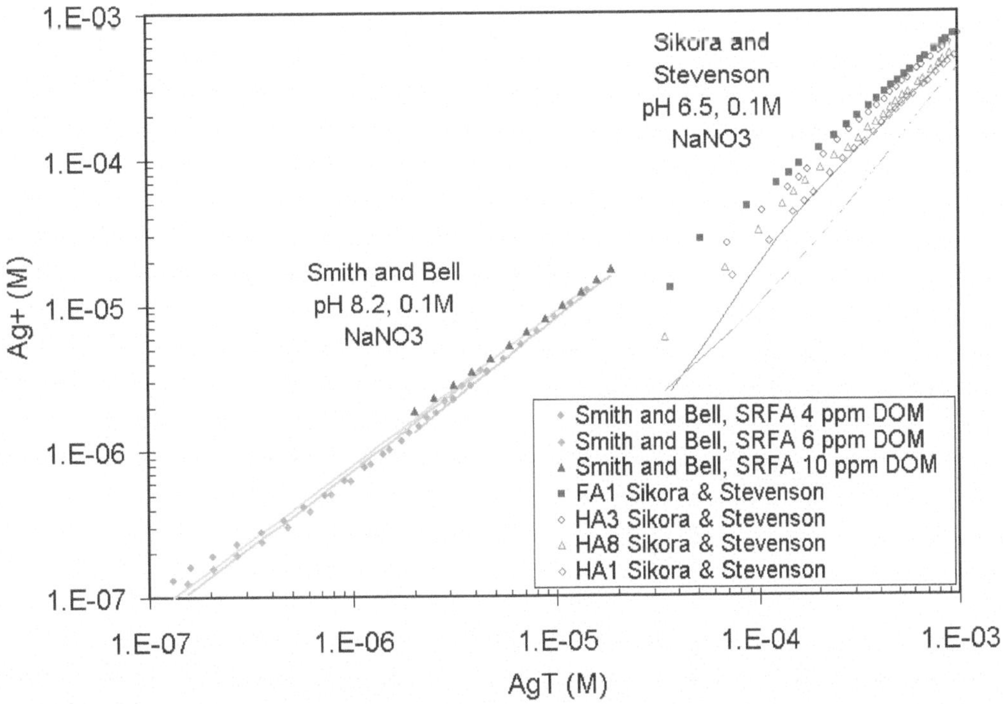

Figure 2-2. Ag Titrations of DOM: (Data: Smith and Bell, 2003; Sikora and Stevenson, 1988). Lines are from BLM simulations showing calibrated weak NOM Interactions based on the distribution of sites in WHAM.

2.4 Summary

The silver BLM, like other BLM versions, relies on characterization of silver speciation in order to explain changes in bioavailability with changes in aquatic chemical environments. Relevant chemical interactions include both inorganic and organic complexation reactions, however, the relative strength of these interactions suggest that whenever reduced sulfur (such as measurable by CRS) is present, interactions with organic matter will dominate silver speciation. Strong groups are represented by chemical moieties that include reduced sulfur (S(II)) including simple sulfides, which may be stabilized against oxidation in metal-sulfide clusters. The binding sites associated with carboxylic and phenolic groups represented in WHAM are relatively unimportant for silver, although they can be calibrated to provide good agreement with measured silver speciation at high Ag/NOM ratios.

The chemical speciation analyses were generally set in accordance with the results presented above and held fixed in the remaining analyses of toxicity data. However, this process was not strictly "feed-forward", as the concentration of strong NOM binding sites was adjusted in light of the overall toxicity response and, specifically, the response of the predicted LC50 values to varying DOC. Once this was set, the chemical calibration was consistent for all of the BLM predictions of LC50 values.

CHAPTER 3.0

TOXICITY TESTS IN MODIFIED LAB WATERS

3.1 Introduction

As mentioned in Chapter 1.0, the SAB recommended that the Ag BLM be extended in its applicability to a broader spectrum of water quality conditions and aquatic organisms. Toxicity tests with *C. dubia* were therefore conducted in lab waters in which the water quality was systematically varied to achieve a wide range of test conditions (i.e., variations in Ca^{2+}, Mg^{2+}, Na^{+}, Cl^{-}, pH, alkalinity, DOC and sulfide). *C. dubia* was used for testing purposes because it is one of the most Ag-sensitive organisms in the EPA WQC toxicity database, and a previous Ag BLM calibration for *C. dubia* had not yet been developed. This chapter reviews the results from these toxicity studies and the application of the BLM to these data. A more detailed description of the methods and results are provided in a supporting manuscript (Naddy et al., 2008). The application of the BLM to Ag toxicity data for *P. promelas* (from experimental work performed prior to the start of this WERF study) is also reviewed as a prelude to the discussion of the natural water test results to be reviewed in Chapter 4.0. The application of the BLM to a more extensive *P. promelas* dataset and results of related physiological test with *P. promelas*, are described in more detail in two project-related manuscripts (Bielmyer et al., 2007 and Wu et al., 2008).

3.1.1 BLM Calibration Approach

The approach to calibration of the chemical interactions in the BLM was previously reviewed in Chapter 2.0. As described, the chemistry calibration resulted in the evaluation of a range of concentrations of strong OM binding sites for Ag which yielded comparable fits of the Ag chemistry data. This led to a subsequent iterative refinement of the Ag BLM chemistry parameterization that was guided, in part, by the ability of the BLM to reproduce the toxicity data. The BLM parameters that control biological interactions were originally evaluated on the basis of gill silver accumulation data for fish (Janes and Playle, 1995) in combination with the achievement of a concurrent fit of the available laboratory toxicity data (see Wu et al., 2008). The fit of the toxicity data was achieved by first evaluating the BL:Ag concentrations for the individual toxicity tests for a given species by running the BLM in speciation mode. When doing so, the dissolved Ag concentration for the individual test waters was set to the measured LC50 for each test and the balance of the water chemistry was defined by the measured water quality characteristics. The lethal accumulation level at the biotic ligand at 50% mortality (i.e., the LA50), a BLM parameter that is adjusted to account for differences in species (and/or age class) sensitivity, was set equal to the average of the BL:Ag values that were calculated for each of the individual tests for a given species (or age class). Given that reasonable consistency in LA50 values that were evaluated over the wide range of exposure conditions tested, it was not considered necessary to log-transform the LA50s prior to evaluating the species-specific LA50s (median and average values were quite similar, within 10-15% of each other for each species). Summaries of the biotic ligand binding constants (BL:Ag log K) and LA50 values are provided in Table 3-1 and 3-2, respectively. This base case set of BLM parameter values will be used for all calculations to be presented, with the exception of a limited number of sensitivity evaluations that will be characterized in detail as they are discussed.

Table 3-1. Summary of Biotic Ligand Binding Constants for Model with NOM Strong Binding Site Density = 2 nmol/mg DOC.

BLM Input Parameter	Log K
BL-Ag	6.8
BL-AgCl	9.7
BL-Ca	1.7
BL-H	3.7
BL-Na	2.3

Toxic species: BL-Ag + BL-AgCl

Table 3-2. Summary of Biotic Ligand Lethal Accumulation Levels for Model with NOM Strong Binding Site Density = 2 nmol/mg DOC.

Aquatic Species	LA50 (nM/g wet weight)	
C. dubia	0.16	
D. magna	0.65	
O. mykiss	10.70	
P. promelas:	<u>IS < 0.01 M</u>	<u>IS > .01 M</u>
4-day old:	2.50	2.50
8-day old:	LA50 = 723.443 x IS - 1.293	5.94
28-day old:	LA50 = 723.443 x IS + 1.375	8.61
42-day old:	LA50 = 723.443 x IS + 6.518	13.75

3.1.2 BLM Version

The version of the Ag BLM that was available at the initiation of the WERF project (Version A005) had been calibrated to toxicity data obtained with modified lab waters and with a limited number of test results with natural waters. That preliminary version of the BLM has been further refined, in accordance with the procedure outlined above, on the basis of laboratory water and natural water toxicity data obtained pursuant to the WERF project. These refinements had only a very limited effect on the lab water results that are reviewed in this chapter. The results to be presented in this chapter were obtained with the current version of the Ag BLM (Version 2.2.4).

3.2 *Ceriodaphnia dubia* Test Results

The methods employed in the toxicity tests with *C. dubia* are described by Naddy and co-workers (Naddy et al., 2008). Briefly, the methods conformed to standard procedures that are followed for the toxicity data to be acceptable for WQC development purposes (ASTM 2001; U.S. EPA, 2002a). The standard reference waters that were used in the toxicity tests were created from one or the other of two alternative base waters, either Milli-Q® grade laboratory water (pH = 8.2, Ca^{2+} = 24 mg/L, Mg^{2+} = 8.0 mg/L, and total hardness = 100 mg/L as $CaCO_3$, total alkalinity = 100 mg/L as $CaCO_3$, Na^+ = 39 mg/L, Cl^- = 4.3 mg/L, DOC = 0.8 mg/L) or Horsetooth Reservoir water (used in selected studies; typically pH = 8.0, Ca^{2+} = 25 mg/L as $CaCO_3$, Mg^{2+} = 6.5 mg/L as $CaCO_3$, and total hardness = 90 mg/L as $CaCO_3$, total alkalinity = 100 mg/L as $CaCO_3$, Na^+ = 16 mg/L, Cl^- = 2.4 mg/L, DOC = 3.3 mg/L). The characteristics of these base waters were systematically varied by the addition of reagent-grade salts ($NaHCO_3$, $CaSO_4$ and $MgSO_4$, as well as KCl in the case of the Milli-Q water) and/or concentrates of natural organic matter (NOM), to achieve environmentally relevant concentration ranges of various water quality characteristics. Pre-equilibration times between the silver and the test water, prior to the addition of the test organisms, were typically less than one hour. It is not expected that this would have had a bearing on most of the test results, since most waters were low in organic matter content. Analytical results for these tests are summarized in Table B-1 in Appendix B. Typically the concentrations of only a limited number of constituents were varied at any one time in an effort to isolate the response of the test organism to changes in concentration of specific water quality characteristics. While the test matrix focused on those constituents commonly thought to affect Ag bioavailability and toxicity (e.g., Na^+, Cl^-, etc.), it was necessary to vary more than one ion at a time because either the cation or the anion that compose the neutral salts that were added could potentially affect Ag bioavailability. The toxicity data provide a measure of the integrated effects of the water quality constituents on Ag bioavailability, and when multiple effects occur simultaneously it can be difficult to know from toxicity data alone how those effects could be attributed to single water quality variables. The BLM, however, can be used to help interpret the results in a more quantitative and chemical-specific level than would otherwise have been possible.

The *C. dubia* 48-hour dissolved Ag LC50 was not significantly affected by changes in hardness over a total hardness range of approximately 100 to 600 mg/L as $CaCO_3$ (Naddy et al., 2008). This is in contrast to the 1987 draft Ag WQC which included hardness as the only water quality factor that affects bioavailability (U.S. EPA, 1987). The hardness relationship in the Ag WQC document was based in part on the juvenile rainbow trout test results of Lemke (1981).

However the apparent hardness response that was reported by Lemke (1981) was later attributed to the chloride, which increased with increasing hardness in those experiments, because the formation of the AgCl complex would be expected to reduce Ag bioavailability (Hogstrand et al., 1996). The test results with *C. dubia* in low DOC waters therefore seem to be in accordance with other relatively recent test results that indicate a limited effect of hardness on the toxicity of Ag to aquatic organisms (Bury et al., 1999a; Karen et al., 1999).

The hardness tests also considered the effect of increasing alkalinity on Ag toxicity, because the suite of conditions tested included the addition of varying amounts of $NaHCO_3$ (alkalinity of 50, 100, and 200 mg/L as $CaCO_3$). Although the effect of alkalinity on LC50 was not statistically significant in these tests, a qualitative increase in LC50 was observed. An analysis of these results using the BLM showed that the increase in alkalinity itself should not be expected to alter either Ag speciation or Ag toxicity, but that an increase in toxic effect levels of about 20% should occur as a result of the increase in Na^+ that occurred in these tests. The increase in Na^+ was a result of the addition of $NaHCO_3$ to adjust the alkalinity. The qualitative increase in Ag effect levels that was observed is therefore entirely explained by the changes in Na^+ ion concentration, rather than from the increases in Ca^{2+} or alkalinity.

The effect of Na^+ on silver bioavailability was further investigated in toxicity tests where salts such as Na_2SO_4, $NaHCO_3$ or NaCl were added to the reference water. The change in Ag toxicity as a function of Na^+, as measured by the 48-hour Ag LC50, is shown on Figure 3-1. The unfilled symbols represent data where either HCO_3^- (squares) or $SO_4^=$ (circles) were added as the counter ion, along with the added Na^+. Since it is known from chemical thermodynamics that neither HCO_3^- nor $SO_4^=$ will form complexes with Ag, the observed response (unfilled symbols) can be attributed to the changing Na^+ concentration rather than to a decrease in Ag bioavailability due to complexation. These data indicate a clear benefit of increasing the Na^+ concentration from 2.5 to 21 mg/L (by adding $NaHCO_3$) and to 160 mg/L (by adding Na_2SO_4). When the BLM was used to simulate these test results, by varying Na^+ in association with the average chemistry of the test water, a slightly reduced response was obtained (dashed line). The BLM accounts for the response via the competition of Na^+ with Ag^+ for binding to the biotic ligand (BL), such that increasing Na^+ leads to the need for a higher dissolved Ag concentration for the Ag:BL concentration to equal the LA50. Any other physiological effects of Na^+, such as a beneficial effect of increasing Na^+ on Na^+ uptake kinetics, are not explicitly represented by the BLM. To the degree such effects are operative, they are implicitly included in the BLM via the response associated with Na^+ as a competing cation.

The second set of test results where Na^+ was varied included the addition of Cl^- as the counter ion (filled diamonds on Figure 3-1; base water ~38 mg/L Na^+ and 4 mg/L Cl^-). At the lower Na^+ levels tested of 38 and 48 mg/L Na^+ (and 4.4 and 20 mg/L Cl^-) the LC50 values were slightly lower than the LC50 values that had been measured in earlier tests where $NaHCO_3$ had been added (compare the filled diamonds versus the unfilled squares). There was a slight increase in the LC50 with increasing NaCl in these two tests, from 0.25 to 0.33 µg/L. At the highest NaCl treatment level tested (150 mg/L Na^+ and 291 mg/L Cl^-, by adding NaCl) the LC50 increased to 1.62 µg/L, more than a factor of 6 increase relative to the lowest NaCl treatment level. The BLM simulates a somewhat weaker response to the addition of NaCl than is indicated by these data, about a factor of 3 change overall, but a distinct increase in LC50 is evident (solid line). The interpretation of these test results is complicated because both Na^+ and Cl^- have the potential to affect the bioavailability and toxicity of Ag. However, in the context of the BLM, the contributions of Na^+ and Cl^- are readily differentiated. The effect of Na^+ is explained by its role as a competing cation. That is, the increase in Na^+ reduces the degree of Ag^+ interaction at the site of action of toxicity (i.e., at the biotic ligand), thereby providing a protective benefit.

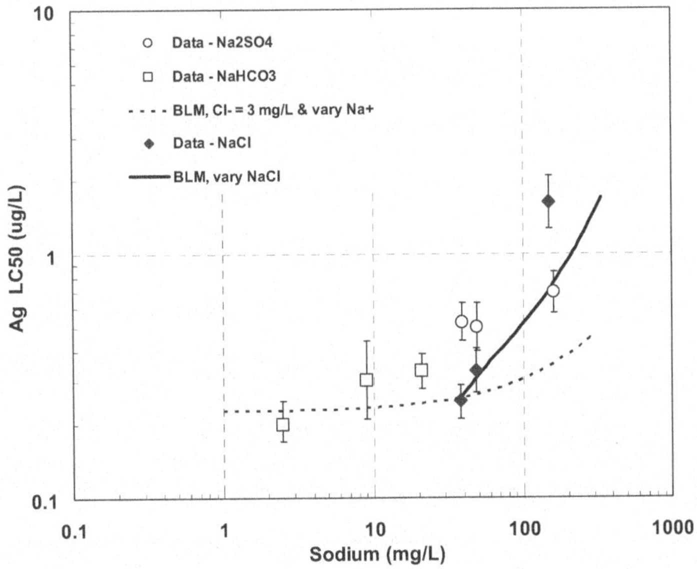

Figure 3-1. Comparison of Measured (unfilled and filled plot symbols, ± 95% CL) and BLM-predicted (dashed and solid lines) Responses of *C. dubia* LC50 to Na^+ and Cl^- (Data: Naddy et al., 2008).

Alternatively, Cl^- contributes to the protective response via the decrease in Ag^+ and the concomitant formation of AgCl (a chloro-complex of Ag that is considered to be less available for uptake than Ag^+), and $AgCl_2^-$ (a Ag complex that is presumed to be non-bioavailable), changes that accompany the increase in Cl^-. Together, these factors lead to a clear beneficial effect associated with the increase in NaCl. The BLM simulation results attribute about 20% of the increase in LC50 to the competitive effect of increasing Na^+ and about 80% of the increase to the formation of the inorganic chloro-complexes.

It is clear from Figure 3-1 that the protective effect of Cl^- is the reason that Na^+ and Cl^- together have a greater effect on Ag toxicity than when Na^+ concentrations are increased in association with either HCO_3^- (squares) or $SO_4^=$. Other studies, however, have shown that the protective benefit of Cl^- against Ag toxicity to a variety of aquatic organisms is limited at best (see Bielmyer et al., 2007 for a review). The apparent inconsistency with the results of Figure 3-1 may in part be explained by the maximum Cl^- concentration used in the test water in most other studies, typically less than about 50 - 55 mg/L. Simulations with the BLM indicate that a relatively limited increase in LC50, on the order of 50%, would be expected over this range of Cl^- levels. An increase of this magnitude would be difficult to detect with a standard toxicity test design and normal test variability. As discussed in Chapter 4.0, while environmental Cl^- levels are typically less than 50-60 mg/L (Davies and DeWiest, 1966), higher levels may be encountered, such as in an effluent-impacted water body or within the transition zone between a freshwater stream or river and an estuary.[3] The protective effect of Cl is discussed in further detail in Chapter 5.0.

---

[3] The chloride level of Desjardins Canal, an effluent-impacted natural water, was 155 mg/L.

With regard to pH, the acidity of the water did not appear to have a consistent effect on the toxicity of Ag to aquatic organisms, at least in low DOC waters. The pH was decreased by pumping 2% CO_2 gas into a sealed test vessel. There was a qualitative increase in the Ag effect level with decreasing pH in one set of tests (pH 8.3 and 7.2) and a qualitative decrease in the other set of tests (pH = 8.3 and 7.5). Simulations with the BLM indicate that pH should have a negligible effect on Ag speciation in low DOC waters over an environmentally relevant range of 5.5 to 9.0. Thus the absence of a consistent response could reasonably be expected. On the other hand, as discussed in Chapter 2.0, pH is potentially more important at elevated DOC levels since both H^+ and Ag^+ compete for binding sites associated with natural organic matter. Thus, Ag complexation by DOC tends to be relatively ineffective at low pH, because the binding sites have an increased tendency to be occupied by H^+ ions (i.e., the DOC binding sites are protonated).

The DOC tests with *C. dubia* were performed over a relatively narrow pH range of 7.5 to 7.9 and with DOC (Dundas WWTP NOM) varying from 1.6 mg/L (reference water) to between 3.5 and 10.5 mg/L (NOM-amended waters). The LC50 values increased nearly 10-fold, from 0.58 to 5.39 µg/L, as the DOC increased. This increase actually reflected the effect of increasing DOC as well as increasing sulfide (CRS = 33-109 nM in the NOM-amended waters). As discussed in further detail in Chapter 4.0, preliminary BLM simulations suggested that the measured increase may still have been understated due to the use of a relatively short time for equilibration of the added silver with the NOM and CRS of the test water prior to the addition of the organisms.

3.3 *P. promelas* Test Results

Toxicity tests with *P. promelas* exposed to Ag in systematically modified lab waters were not included in the WERF program. Rather, the initial calibration of the Ag BLM to *P. promelas* data relied upon results from earlier studies where tests of that type were performed (e.g., Erickson et al., 1998; Karen et al., 1999; Bury et al., 1999a; see Wu et al., 2008 for a summary of BLM-related analyses of these data). However, because validation of the Ag BLM with toxicity test results obtained with unmodified natural water samples (Chapter 4.0) were included in the WERF investigations, it is useful to describe the status of the current Ag BLM calibration for *P. promelas*.

To begin, consider the comparison of LC50 data (Karen et al., 1999) to BLM simulation results (Wu et al., 2008) shown on Figure 3-2. These data were obtained with relatively sensitive 4-day old *P. promelas* that had been acclimated to water having a hardness equivalent to that of the exposure water for a period of 3 days. The dataset was also minimally influenced by low ionic strength conditions, a factor that complicates the interpretation of the test results with older *P. promelas* that have been analyzed (refer to Chapter 4.0 and to Wu et al., 2008). Though not well documented, it is understood that these tests were performed with a relatively short equilibration time, on the order of a few hours or less (Stephen Klaine, personal communication). Figure 3-2 displays test results for the four nominal chloride levels tested (indicated above each graph: 3, 20, 40 and 60 mg/L, added as KCl, with 42 mg/L Na^+ in the base water). Each panel compares the measured LC50 values (unfilled bars) to LC50 values simulated with the BLM (filled bars), for each of 3 DOC levels (1.2, 5 and 10 mg/L DOC, added as Aldrich humic acid, AHA). Finally, the range indicated with each bar corresponds to the variation in LC50 that was observed at the three hardness levels tested (55, 110 and 220 mg/L as $CaCO_3$, added as $CaSO_4$ and $MgSO_4$ at a 1:1 ratio on a mass basis).

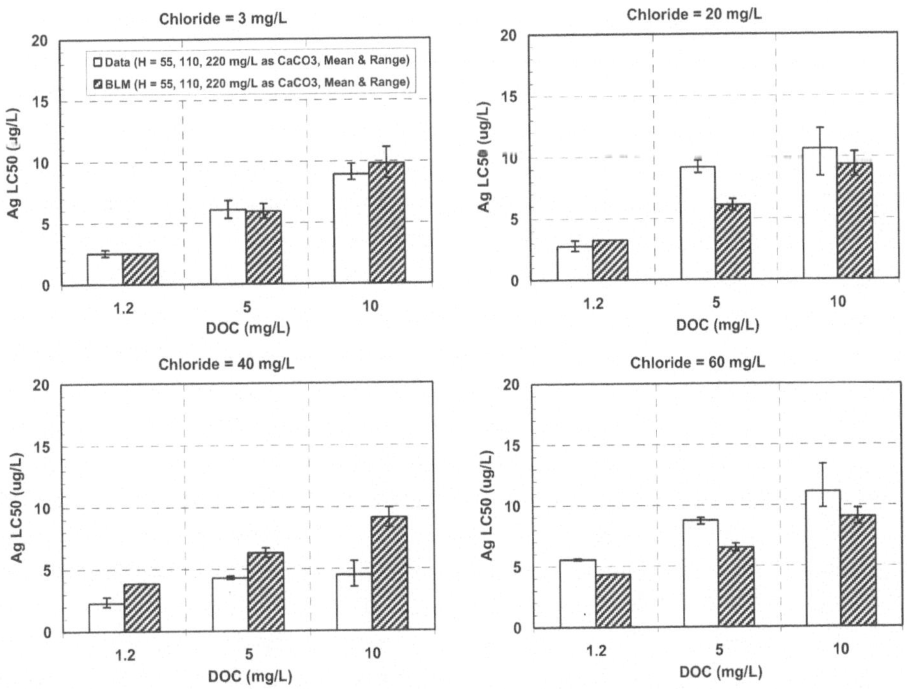

Figure 3-2. Comparison of Measured (unfilled bars) and BLM-Predicted (filled bars, without any effect of IS on LA50). Responses of 4-day old *P. promelas* LC50 to Cl^- (added as KCL), DOC (added as AHA) and Hardness (added as $CaSO_4$ and $MgSO_4$, on a 1:1 mass basis) (Data: Karen et al., 1999).

The overall agreement between the model and data of Figure 3-2 is quite good. The model reproduces the overall trends exhibited by the data without evidence of a consistent bias. The increase in Ag LC50 on each panel reflects the response to DOC. The impact of increasing DOC concentration is greatest for the low chloride treatment, a subset of the results that the model simulates quite well. The agreement between model and data is also reasonably good at the three higher Cl treatment levels, though the model produces a slight overestimation of the observed LC50 response to DOC at 40 mg/L Cl^- and an underestimation of the LC50 response to DOC at 20 and 60 mg/L Cl^-. Further consideration of these comparisons in regard to the potential effect of equilibration time is discussed in combination with the other test results in Chapter 5.0.

Comparison of the results on each of the four graphs of Figure 3-2 shows the response of the 4-day old *P. promelas* LC50 to Cl^- level. Upon initial inspection, it is apparent that the measured LC50 progression as chloride increases is somewhat variable, first increasing, then decreasing and then increases again as Cl^- is increased (nominally) from 3 to 20 to 40 to 60 mg/L. Karen et al (1999) noted the "somewhat anomalous results" for the *P. promelas* tests at 40 mg/L Cl^-, but did not offer an explanation to account for this response. While it could reasonably be posited that the intermediate shifts (higher LC50 values at 20 mg/L Cl^- than at 40 mg/L Cl^-) reflect the inherent variability of the toxicity tests, additional information to support this is lacking. The response is not entirely erratic either, since the observed LC50 values decrease between 20 and 40 mg/L Cl^- at all three DOC treatment levels. However, a defensible rationale that would support the modification of the Ag BLM to reproduce this type of response was not apparent and, given that the response was judged to be an atypical one, no attempt was made to implement any such modification.

In spite of the preceding unexplained variations in LC50 with Cl^-, there are also some overall patterns that emerge from the data, responses that the BLM tends to reproduce. First, at each DOC treatment level there is an overall increase in measured LC50 when comparisons are made between the lowest and highest Cl^- levels (i.e., between 3 and 60 mg/L). Both the data and the BLM exhibit a reasonably consistent trend of increasing LC50 with increasing Cl^- at the lowest 1.2 mg/L DOC level. The protective effect of Cl^- is slight in either case, with the measured LC50 increasing about 2.2-fold over the range of Cl^- levels while the predicted LC50 increases slightly less, by 1.73-fold. The observed increase in Ag LC50 as Cl^- increased tended to be less pronounced at the highest DOC level (a non-significant 1.25-fold increase was measured). The BLM results, which reflect binding of both Ag^+ and AgCl to the biotic ligand and contributing to the LA50, also exhibit a systematic if somewhat variable effect on LC50 with increasing chloride, across all DOC levels. As is the case for the data, the predicted response is most clearly evident at the lowest DOC level and becomes less pronounced as DOC increases. Interestingly, at the highest DOC level, the BLM actually predicts a slight decrease (0.92-fold) in Ag LC50 as Cl^- increases compared to a non significant 1.25-fold increase in the LC50 data. A response of this nature will be predicted in situations where the level of organic matter (OM) is high enough for an increase in Cl^- to increase the AgCl fraction, a bioavailable form, at the same time that the Ag-OM fraction, a complex that is not considered to be bioavailable, decreases. Unfortunately, it is not possible to confirm that this admittedly subtle BLM response is correct on the basis of the available data, despite that fact that it appears to be a reasonable one. This response and others of a similar nature are discussed in further detail in section 5.2.5.

Other tests with fish have shown that Cl^- appeared to provide protection against Ag toxicity to *O. mykiss*, but not to *P. promelas* or *Anguilla anguilla* (American eel) (Bury et al., 1999a; Grosell et al., 2000). Given these apparent species-specific differences, this response was investigated further in a series of toxicological and physiological studies using *P. promelas* (4 days old), *Danio rerio* and *Fundulus heteroclitus* (Bielmyer et al., 2008). The Cl^- level in the toxicity tests with *D. rerio* and *P. promelas* was increased 1 mM (from 0.01 mM KCl in the reference water, with Na^+ = 0.05 mM = 1.15 mg/L at both Cl^- levels) by the addition of KCl. All tests were performed with a 24-hour pre-equilibration time. The toxicity test yielded a non-significant increase in LC50 of about 10% for *D. rerio* (from 10.3 to 11.3 µg/L) and 15% for *P. promelas* (from 2.32 to 2.67). A decrease in LC50 (once again, not statistically significant) was also observed for the third species (*F. heteroclitus*), regardless of whether the Cl^- was adjusted by adding 1 mM of KCl or $MgCl_2$ (the Ag LC50 decreased from 3.88 µg/L in the reference water to 1.2 µg/L and 3.2 µg/L, respectively, at 1 mM Cl^-). These results are therefore consistent with the earlier LC50 results for *P. promelas* in which the protective benefit of Cl^- was not statistically significant (Bury et al., 1999a). However, as discussed in Chapter 5.0, use of K^+ as the cation that was added along with Cl^- in some of the preceding experiments may be confounding the interpretation of these test results. This is because K^+ has been shown to have ionoregulatory effects on some aquatic organisms and, in doing so, it increases their sensitivity to ionoregulatory toxicants such as Cu and Ag (e.g., Wilcox and Dietz, 1995; Erickson et al., 1996).

The physiological tests were conducted to further the current understanding of the importance of ionoregulation and how it affects the responses of aquatic organisms to silver. Specifically, they were designed to investigate if species-specific differences in the protective effect of Cl^- were in some way related to differences in the rates of chloride uptake and/or differences in the mechanism of uptake by different test species (Bielmyer et al., 2008). While these tests provide further insight into the potential causes of such differences, they were not able to provide a complete explanation of the observed species-specific differences in sensitivity to Ag. The implications of these results to the development of an acute Ag BLM are discussed in further detail in Chapter 5.0.

The variation in Ag LC50 that was associated with varying the hardness from 55 to 220 mg/L as $CaCO_3$, indicated by the ranges shown with each bar on Figure 3-2, was generally small and often inconsistent with respect to a systematic response. Any differences related to changes in the hardness cations could not be readily discerned given the inherent variability of the toxicity tests. Interestingly, while the model results also varied only slightly, the BLM predicted that the LC50 could either increase very slightly with increasing hardness, for low DOC levels (a negligibly small hardness response) or decrease with increasing hardness at high DOC levels. The explanation for this is analogous to the explanation for the effect of pH on the LC50, where both low pH (i.e., high $[H^+]$) and high Ca^{2+} levels lead to increased binding of these competing cations to DOC binding sites, thereby reducing the binding of Ag^+ to these sites. The net effect is that high Ca^{2+} levels can reduce the effective complexation capacity of DOC for Ag^+, and this in turn reduces the mitigating effect of DOC complexation on Ag toxicity. A detailed inspection of the fathead minnow toxicity data for selected treatment suggests that this response may actually be occurring. For example, at a Cl^- of 20 mg/L and hardness levels of 55, 110, 222 mg/L as $CaCO_3$, the respective Ag LC50 values are 9.7, 9.2 and 8.7 µg/L at a nominal DOC of 5 mg/L, and 12.3, 11.2 and 8.4 µg/L at a nominal DOC of 10 mg/L. This predicted response is less definitive for more sensitive organisms because the hardness cations are competing with Ag^+ for relatively strong Ag-NOM-binding sites at these lower, more environmentally relevant Ag effect levels. Unfortunately, the background variability that is associated with the overall dataset makes it difficult to conclusively establish the validity of this predicted cause-effect relationship. It would be of interest to confirm these BLM predictions of this response with further testing. However, any such demonstration would need to include a high degree of replication of the toxicity tests in order to be able to quantify the significance of any interactions that were observed.

The analysis of the *P. promelas* data of Figure 3-2, in combination with several other datasets in which older, larger individuals were tested (Erickson et al., 1998 and Karen et al., 1999 with 28-d fish; Bury et al., 1999a, with > 42-day old fish; Bielmyer et al., 2007 with 8-d old fish) provided an indication that the sensitivity of *P. promelas* to Ag decreases with increasing age or size. Brix (unpublished data) and Bielmyer et al. (2007) each tested different age fish under identical conditions and confirmed this general relationship of decreasing sensitivity with increasing size (Figure 3-3, inset) over a range in age of 1 to 42 days (the LC50 values by Bielmyer and coworkers for 1- and 4-day old fish were essentially the same). There appears to be at least one exception to these datasets, as the LC50 values reported by Van Genderen et al. (2003) for <24-hour old *P. promelas* were higher than the preceding results for 1-day and 4-day old fish. This exception, which could not be explained by differences in water chemistry or any other known artifact related to the experimental conditions, remains unaccounted for at this time.

Grosell et al. (2002) had previously shown that the sensitivity of aquatic organisms to Ag is directly related to the turnover rate of sodium (the mass of sodium in the exchangeable sodium pool divided by the rate of loss of sodium from the body), a physiological trait that is inversely related to size. That a decrease in Ag LC50 should be inversely related to the sodium turnover rate is intuitively reasonable, provided that the lethal threshold of $\%Na^+$ loss is comparable for any size fish. This conclusion arises from the follow line of reasoning. Under conditions of Na^+

homeostasis, the uptake of Na^+ must be sufficient to balance Na^+ efflux (including branchial and renal losses), such that uptake = efflux. For a fixed Ag^+ concentration (the same reasoning applies if other Ag species are thought to be bioavailable) and a fixed % inhibition of Na^+ uptake (e.g., 50%), a fish having a high Na^+ uptake (and efflux) rate will experience a net rate of loss of Na^+ (Na^+ uptake minus efflux) that is faster than a fish having a low uptake rate. Thus, over a fixed time period, it will lose more plasma Na^+. It follows that for a fixed time period (e.g., 96 hours), a lower Ag^+ concentration and a lower % inhibition of Na^+ uptake is needed for a fish with a high Na^+ turnover rate (a small fish) than one with a low Na^+ turnover rate (a relatively large fish) to lose the critical amount of plasma Na^+ that will result in death. Thus, sensitivity to silver, as indicated by the Ag LC50, should be inversely related to size.

Measurements of Na^+ and Cl^- turnover rates of different age/size *P. promelas* conformed to the aforementioned relationship, with the smaller more sensitive individuals tending to have the higher turnover rate (Figure 3-3; Bielmyer et al., 2007). Aside from the slight leveling off of the response for the smallest age classes (the 1- and 4-day fish), these results provide direct support for the explanation that the change in sensitivity to Ag with size is related to the change in turnover rates of Na^+ and Cl^- with size. While approaches for simulating this effect of Na^+ turnover rate on sensitivity have been proposed (Grosell et al., 2002; Paquin et al., 2002b), the current version of the acute BLM does not explicitly represent Na^+ turnover rate. A somewhat simpler and more readily implemented approach that has been adopted to account for species or size sensitivity is to adjust the LA50. Thus, 4-, 8-, 28- and >42-day old *P. promelas* are assigned LA50 values of 2.50, 5.94, 8.61 and 13.8 nmol/g_w, respectively (with adjustment of the LA50 for the older fish as summarized in Table 3-2). The result is that, for a fixed set of water quality characteristics, the predicted *P. promelas* LC50 will decrease with a decrease in their age or size.

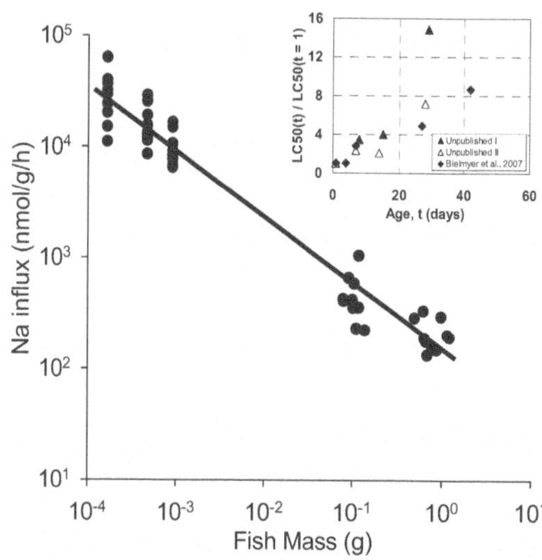

Figure 3-3. Variation of *P. promelas* Na^+ Uptake Rate with Size and Sensitivity to Ag with Age. (Data: Bielmyer et al., 2007).

3.4 Analysis of the Proof of Principle Tests Results and Protection by Sulfide

Because silver is known to be toxic at relatively low levels in lab waters (as compared to metals such as copper and zinc) and ambient environmental levels are also known to be quite low, dissolved ligands that are present at low concentrations in aquatic systems may still have an important effect on silver speciation and, potentially, its bioavailability and toxicity to aquatic life (Campbell et al., 2002). Sulfide, which is typically found at sub-micromolar concentrations in natural waters (Kuwabara and Luther, 1993; Kramer et al., 1999; Rozan et al., 2000; Bowles et al., 2002a,b), is one such ligand that would be expected to control silver speciation (Bell and Kramer, 1999; Rozan and Luther, 2002).

A number of datasets in which organisms were exposed to Ag in the presence of sulfide were analyzed with the silver BLM as part of the WERF project. The purpose was to determine the best way to represent the protective effects of sulfide on silver toxicity. Two of the key datasets that were used involved tests with *Daphnia magna* (Bianchini et al., 2002), and *Oncorhynchus mykiss* (rainbow trout; Mann et al., 2004). Since these experiments were specifically designed to test the validity of the previously presumed protective effect of sulfide on silver toxicity, they are collectively referred to as the Proof of Principle (POP) experiments. This subsection describes the results of these experiments and the application of the BLM to assess the protective effect of sulfide against toxicity due to silver.

3.4.1 Proof of Principle Experiment with *D. magna*

D. magna neonates were used in a series of 48-hour static exposures designed to test the protective effect of sulfide on silver toxicity (Bianchini et al., 2002). The exposures were performed in synthetic moderately hard water consisting of reverse osmosis-purified (RO-purified) water that was reconstituted to the following composition: pH = 8.2, hardness = 115 mg/L as $CaCO_3$ with Ca/Mg molar ratio of 6.67, Na = 14 mg/L and Cl = 21 mg/L. The oxic exposures were conducted in the absence of sulfide, in the presence of 25 nM sulfide (nominal), and in the presence of 250 nM sulfide (nominal). The sulfide was introduced into the test water in the form of zinc-sulfide clusters. Further details concerning the synthesis of the zinc-sulfide clusters and the chemical system that was used and for these experiments can be found elsewhere (Bowles et al., 2002a,b). Silver (as $AgNO_3$) was added to the test vessels 3 hours before the addition of the organisms. Related tests had indicated that 3 hours was sufficient time for near-equilibrium conditions to be achieved between the added Ag and the inorganic sulfide. Specifically, the dissolved inorganic sulfide concentration (measured with mixed diamine reagent, MDR, which does not measure AgS) in the presence of Ag, was lower than in the absence of Ag and the difference was nearly constant following an initial 3-hour time period (Bowles et al., 2002b). Further, the difference in measured sulfide concentrations was approximately equal to the molar concentration of the added silver, suggesting that the binding of silver to sulfide was occurring at a 1:1 ratio (assuming none of the sulfide bound to silver reacted with mixed diamine reagent, MDR). Bowles et al. (2002a), also showed that reaction of sulfide with MDR, in the presence of Ag, was approaching completion within 30 minutes (i.e., comparable results were obtained after 48 hours). It was inferred from these results that the reaction of Ag with the sulfide was nearly complete within 30 minutes. As a result, the authors provided what was described as a conservative recommendation that a 3-hour pre-equilibration time be used for a system containing Ag and ZnS clusters.

The *D. magna* silver LC50 (based on the total recoverable silver) in the two exposures conducted in the absence of sulfide averaged 0.22 µg/L (2.04 nM) (Figure 3-4, upper panel, control). This may be compared to results with 25 nM of sulfide where the total recoverable Ag LC50 increased to 1.47 µg/L (13.6 nM) (results "with sulfide"). The comparison demonstrates the principle that sulfide has the ability to mitigate silver toxicity. Further, the presence of 250 nM of sulfide provided complete protection against toxicity even in the highest silver treatment in these experiments (2 µg/L or 18.5 nM; results not shown). It should be noted that the dissolved Ag LC50 was considerably lower than the total recoverable Ag LC50 for these experiments. The investigators attributed this finding to sorption of the AgHS complex to the filter pad during sample filtration, a known experimental artifact of using Zn-sulfide clusters that has been documented by Bowles et al. (2002a). In view of this the BLM predictions for these experiments will be compared to the total Ag LC50 results.

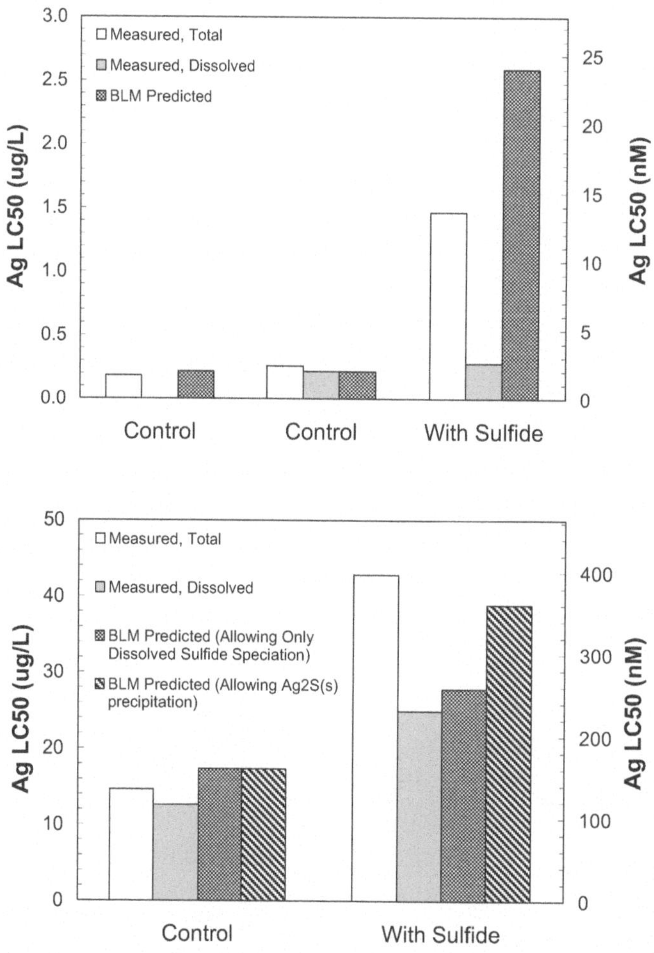

Figure 3-4. Analysis of Proof of Principle Experiment.
(Data: Upper panel, *D. magna*, Bianchini et al., 2002; Lower panel, *O. mykiss*, Mann et al., 2004)

The inorganic sulfide and metal-sulfide interactions shown in Tables 2-2 and 2-3 (Chapter 2.0, Silver Chemistry) were incorporated in the BLM and used to predict LC50 results in the POP experiment for *D. magna*. Direct application of the BLM led to an over-prediction of the lab water LC50. The problem was apparently related to the atypically low LC50 that was measured in the lab water (0.22 µg/L), an LC50 that was lower than several other *D. magna* Ag LC50 values (0.5-1.5 µg/L), including test results previously used in the calibration of the BLM (Nebeker et al., 1983; Elnabarawy et al., 1986; Erickson et al., 1998; Karen et al., 1999). Bianchini and coworkers recognized that the LC50 for the POP organisms was relatively low and hypothesized that this may have been related to the chemical characteristics of the test water (i.e., the low alkalinity of ~100 mg/L as $CaCO_3$ and low DOC ~0.1 mg/L). However, BLM simulation results indicate that these differences in water chemistry were not likely to be sufficient to account for the differences in LC50 values. These simulated results seem reasonable because the alkalinities for the tests used to develop and calibrate the BLM were in most cases lower than the alkalinity of the POP tests and the DOC, while slightly higher, was also low (1.0-1.5 mg/L, when reported). It was also suggested that use of nominal measurements may account for the higher LC50 values but only one of the other studies based the results on nominal concentrations (Elnabarawy et al., 1986). Measured Ag results were used for all of the other studies and the LC50 was still higher than in the POP experiment (i.e., Nebeker et al., 1983; Erickson et al., 1998; Karen et al., 1999). Hence the reason for the low LC50 values for the POP test organisms cannot be readily explained. As a test of the ability of the BLM to predict the protective effect of sulfide, the LA50 was set to achieve consistency with the reference water LC50. When this was done the BLM was able to predict an increased LC50 in the presence of added sulfide of 2.6 µg/L (24.0 nM), a value in reasonably good agreement with the measured total LC50 of 1.47 µg/L (13.6 nM) (Figure 3-4, upper panel, "BLM Predicted", with sulfide). Thus, the BLM response to the presence of sulfide was reproducible. In the exposure with sulfide, silver speciation is dominated by complexation with sulfide, specifically the AgHS complex. Given the high binding constant for AgHS formation, toxicity occurs only after essentially all sulfide is complexed with silver.

3.4.2 Proof of Principle Experiment with Rainbow Trout

The second POP experiment used juvenile rainbow trout (weight <1.0 g) in a series of 96-hour static-renewal exposures with silver (Mann et al., 2004). The purpose was to assess the protective effect of sulfide on silver toxicity to a fish, tests performed at much higher Ag levels than used in the preceding *D. magna* experiments. The exposures were performed in Hamilton (ON, Canada) tap water with the following composition: pH = 7.8-8.0, hardness = 140 mg/L as $CaCO_3$ with Ca/Mg molar ratio of 5, Na = 14 mg/L and Cl = 25 mg/L. Exposures were conducted under oxic conditions in unmodified tap water (no sulfide present), and in the presence of 100 nM (nominal) sulfide. As with the *D. magna* exposures, sulfide was added in the form of zinc-sulfide clusters.

The rainbow trout silver LC50 (based on the total recoverable silver) in the exposures conducted in the absence of sulfide averaged 14.7 µg/L (136 nM) (Figure 3-4, lower panel, control). The LC50 increased to 42.8 µg/L (397 nM) in the presence of 100 nM sulfide, thereby demonstrating the protective effect of sulfide on silver toxicity (results "with sulfide). As with the preceding set of results, because the experiments were performed with laboratory water that was otherwise low in complexing ligands, the protective effect on silver toxicity can be attributed to sulfide. It should also be noted that several of the silver treatments in the presence of sulfide exhibited $Ag_2S_{(s)}$ precipitation, evidenced by the presence of black mucus threads in the test chambers. The possibility of $Ag_2S_{(s)}$ precipitation is stoichiometrically consistent with the toxicity data. That is, the LC50 of 136 nM in the absence of sulfide increases to 397 nM in the presence of 100 nM of sulfide, suggesting that toxicity was exerted only after the 100 nM of sulfide reacted with 200 nM silver forming $Ag_2S_{(s)}$. Once this stoichiometric requirement is accounted for, the excess silver (397-200 = 197 nM Ag) is within 45% of the LC50 measured in the absence of sulfide (136 nM Ag).

The results of the BLM application to the rainbow trout experiments are also shown on the lower panel of Figure 3-4. Applying the previously calibrated rainbow trout silver BLM works well in this case, as seen from the comparison between measured and predicted LC50 values in control water (data and model results for "control"). Toxicity in the exposure conducted in the presence of sulfide was predicted under two scenarios - allowing only dissolved sulfide speciation, and allowing the precipitation of $Ag_2S_{(s)}$. The BLM-predicted LC50 (27.8 µg/L or 258 nM) with only dissolved sulfide speciation tends to under-predict the measured total LC50 (42.8 µg/L = 397 nM) and is slightly higher than the dissolved LC50 and, in either case, is within a factor of two of the measured values. Allowing the $Ag_2S_{(s)}$ precipitate to form results in a more favorable comparison between the measured total LC50 (42.8 µg/L = 397 nM) and predicted total LC50 (38.9 µg/L = 361 nM). In the former case, toxicity occurs only after 100 nM of sulfide is complexed with 100 nM of Ag (the 1:1 stoichiometry AgHS species tends to dominate the speciation). In the case with precipitation, toxicity occurs only after satisfying the stoichiometric requirement of 200 nM of silver to bind with 100 nM of sulfide to form $Ag_2S_{(s)}$.

It is unclear why precipitation would be a factor in the rainbow trout dataset and not in the *D. magna* dataset. Stoichiometric considerations (from comparing the Ag LC50 with the added sulfide) as well as visual observations of a black precipitant in the toxicity exposures seem to indicate that precipitation only occurred in the rainbow trout experiment with sulfide. This is puzzling since a comparison of the ion activity product to the solubility product for $Ag_2S_{(s)}$ suggests that precipitation should have been a factor in both sets of results. One possible reason for this discrepancy may be the higher silver concentrations present in the rainbow trout experiments relative to the *D. magna* experiments and a correspondingly higher degree of oversaturation. Since precipitation seems absent or somewhat unimportant in the *D. magna* exposures, an organism with effect levels that are comparable to the freshwater silver criterion, precipitation could reasonably be excluded from BLM applications for criteria purposes. When handled in this manner, if precipitation were to occur, then the BLM predictions would err on the conservative (i.e., safe) side, by underestimating the LC50.

3.5 Summary

The laboratory toxicity test results, in combination with previously published toxicity test data, served as the basis of the initial calibration of the acute Ag BLM. The results presented in this chapter reflected subsequent refinements to this earlier version of the model. The refinements were introduced in response to the initial application of the model to natural water test data (discussed in Chapter 4.0), an exercise which highlighted a need for the model calibration to be refined to improve its concordance with the overall dataset being used in the BLM development effort. These refinements had only a slight effect on the results of the earlier calibration of the model to lab water data. Pursuant to the BLM simulations and toxicity data, the factors having the largest effect on silver toxicity are NOM, Na^+, Cl^- and CRS. Interestingly, while an increase in the concentration of any of these constituents will usually serve to mitigate

Ag toxicity, the BLM suggests that this is not always the case. Examples of situations where the BLM predicts decreases in effect levels with increasing Cl^- or Ca^{2+} were described, though the responses were slight. The BLM was also successfully applied to the results of the POP experiments. These experimental results had verified the previously unconfirmed belief that complexation of Ag by sulfide would reduce its availability and toxicity to aquatic organisms. It is important for the BLM to be able to simulate this protective effect because relatively low levels of CRS can be expected to have a marked influence on the toxicity of silver in aquatic settings.

CHAPTER 4.0

TOXICITY TESTS IN NATURAL WATERS

4.1 Introduction

Pursuant to their initial review of the Ag and Cu BLM, the EPA SAB recommended that further BLM-related investigations be conducted (U.S. EPA, 2000). The recommendations included: (i) the independent verification of the model with additional datasets, and (ii) assessment of the accuracy of the model under a broader spectrum of water quality conditions and aquatic organisms. The results of investigations that were directed at satisfying these needs for Ag are reviewed in this chapter. These investigations included the completion of Ag toxicity tests with *C. dubia* and *P. promelas* in a variety of natural waters, with concurrent tests in laboratory reference waters, and analysis of these datasets with the Ag BLM. Test results using samples of concentrated natural organic matter (NOM) that were reconstituted in synthetic laboratory waters were also performed with *C. dubia*. As described in this chapter, the initial efforts to simulate these results with an earlier version of the BLM, particularly for *P. promelas*, led to the identification of limitations in the model that warranted further study. Physiological investigations were therefore performed with *P. promelas* to gain an improved understanding of the mechanisms that controlled their response to Ag in the natural waters that were tested. These model testing investigations led to refinements to the earlier calibration in order to enhance the overall performance of the Ag BLM.

4.2 Natural Water Testing and Related Investigations

Data for two sets of natural water samples will be considered. Natural water Dataset 1 consists of *C. dubia* toxicity test results that were obtained with natural water samples and POTW effluent samples from a variety of sites. It also included tests with synthetic lab water that had been adjusted to the hardness of these site waters, or amended with NOM that was reconstituted from NOM concentrates obtained from the site waters. Natural water Dataset 2 consists of eight natural water samples that were obtained from locations throughout North America and included tests with both *C. dubia* and *P. promelas*.

4.2.1 Dataset 1: *C. dubia* in Natural Waters and Water with Reconstituted NOM

The first set of natural water toxicity tests to be reviewed, Dataset 1, was conducted with *C. dubia* exposed to silver in samples from a surface water that was not impacted by a treatment plant discharge (Black River) and two wastewater-impacted surface waters (one sample from the Grand River and two samples from Desjardins Canal, Hamilton, Ontario) (Appendix B, Table B-2). A wastewater effluent sample from the Rochester, New York, WWTP was also tested. The reference water was deionized tap water that was passed through a Millipore unit. Additionally, *C. dubia* toxicity tests were performed with synthetic lab water that was amended with NOM that had been previously concentrated from these same natural water samples. The NOM was concentrated by reverse osmosis (RO) and/or rotary evaporation (RE) techniques and subsequently reconstituted to the DOC concentrations of the original water samples.

The original purpose of the Dataset 1 investigations was to develop a way to extract organic matter from natural water and effluent samples such that it could subsequently be used to test the response of *C. dubia* to Ag in lab water where the DOC was uniformly increased. The toxicity tests compared results obtained with the unfiltered and filtered test waters and with lab waters that were amended with NOM concentrate to achieve a DOC level that was comparable to the original water sample. A total of 22 samples were tested. The chemical characteristics of the filtered natural water samples are summarized in Appendix B, Table B-2. The combined set of natural water and WWTP effluent samples exhibited wide ranges in most dissolved water quality characteristics: pH = 7.2 - 8.3, DOC = 3.2 - 10.9 mg/L (266 - 908 μM), Na^+ = 1.5 - 110 mg/L (65 - 4,780 μM), Cl^- = 2.5 - 221 mg/L (70 - 6,220 μM), Ca^{2+} = 3.5 - 90 mg/L (87.5 - 2,250 μM), Mg^{2+} = 1.3 - 24 mg/L (53 - 988 μM), total hardness = 14 - 314 mg/L as $CaCO_3$ with molar ratios of Ca/Mg = 1.63 - 2.34, K^+ = 2.3 - 13 mg/L (59 - 332 μM) and CRS = 36 - 202 nM.

The *C. dubia* lab water toxicity test results (lightly filled, grey symbols), some of which were presented in Chapter 3.0, the lab water results where the hardness was adjusted to match the natural water samples (i.e., the hardness of the site waters, filled black diamonds) and the results for the site waters themselves (unfilled and filled squares) are compared to the BLM predictions on Figure 4-1. The graph shows the predicted versus measured Ag LC50 for each sample, with the solid diagonal line corresponding to the line of perfect agreement between model and data and the dotted diagonal lines defining the region within which the model predictions are within a factor of two of the data. The first group, the lab water results of Chapter 3.0, exhibit noticeable scatter relative to the range of measured LC50 values (0.17-1.03 μg/L) when displayed in this manner. A more detailed inspection of the individual treatments shows that the BLM-predicted LC50 values covary reasonably well as the measured LC50 increases with increasing sodium and chloride. Conversely, in the cases of pH and hardness, where the data failed to exhibit a systematic variation with water chemistry and the BLM predicts a minimal response, a similar covariation of the results is absent. Calculations with the BLM indicated that the displacement of the LC50 values across different treatment groups (i.e., Na^+, Cl^-, pH and hardness) was not likely due to changes in water chemistry over time. Rather, the differences were within the limits of the quality control charts maintained by the lab. As such, it was concluded that they were simply a reflection of normally encountered variability in unamended lab water LC50 values over time (Naddy et al., 2008).

The second group of results on Figure 4-1, for the low NOM and low CRS lab waters that were adjusted to site water hardness and tested concurrently with the site waters, is actually simulated fairly well by the BLM (filled diamonds). This indicates that the sensitivity of the *C. dubia* that were used for the natural water tests were within the range observed in the earlier investigations with lab water. In contrast to these results, the initial LC50 predictions for the natural water samples themselves were unacceptably high (unfilled squares). The problem seemed to be related to the relatively high CRS concentrations for many of these samples, since this led to calculated concentrations of complexed non-bioavailable silver that were considerably higher than the measured LC50 values. Given that these predictions seemed to be a reasonable expectation (i.e., the POP experiments demonstrated the protective effect of CRS), a variety of exploratory investigations were pursued in an effort to explain this inconsistency between model and data. These analyses included consideration of alternative stoichiometric relationships for silver and sulfide [Ag_2S, AgHS and $Ag(HS)_2^-$], competition between Ag^+ and other cations for sulfide (thereby reducing the effective amount of sulfide that was available to complex Ag^+), and non-equilibrium conditions that reflected the kinetics of Ag^+ complexation reactions. While the first two possibilities failed to account for the deviations from the observations as outlined below, complexation kinetics did provide a plausible and reasonably consistent explanation of the results.

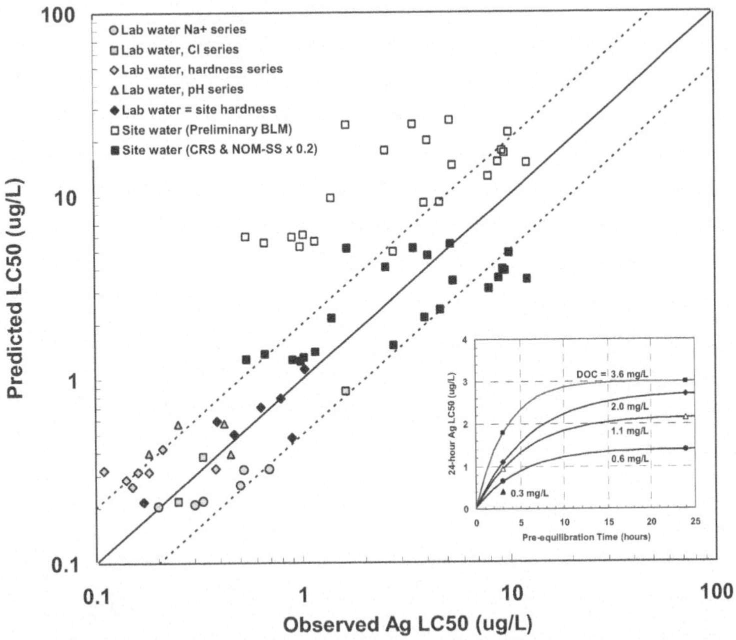

Figure 4-1. Predicted versus Measured Natural Water (Dataset 1) and Lab Water LC50 Results for *C. dubia* and *D. magna* (Inset). (Data: Appendix B, Table B-2; inset after Glover et al., 2005a)

The conventional wisdom prior to the time that the natural water toxicity tests of Figure 4-1 were performed was that complexation of Ag^+ by sulfide proceeded very quickly, over a time scale of minutes (e.g., Adams and Kramer, 1998). However, investigations that were being conducted in parallel with the WERF studies were yielding test results that showed that this was not necessarily the case for all situations, particularly in waters with elevated levels of organic matter (Glover et al., 2005a; Naddy et al., 2007). This point is illustrated on the inset of Figure 4-1. Here, 24 hour LC50 values for *D. magna* increase with increasing organic matter (DOC = 0.6, 1.1, 2.0 and 3.6 mg/L, from the lowest to highest curve, respectively) and, more to the point, also increase (> 2-fold) with respect to the time allowed for pre-equilibration of the added Ag with the water sample prior to initiation of the toxicity test (3 hours versus 24 hours). Preliminary analysis of the data of Glover et al. (2005a) indicate that near-equilibrium conditions (as indicated by the biological response to silver) are achieved in <24 hours in accordance with a kinetic rate constant of about 20-30% per hour (i.e., the curves on the inset correspond to first order rate constants of 0.21, 0.19, 0.17 and 0.31 per hour for DOC amendments of 0.6, 1.1, 2.0, and 3.6 mg/L, respectively).

Most of the *C. dubia* tests performed with the natural water samples of Dataset 1 (Figure 4-1) were completed during 2001 and 2002. The experimental protocol that was followed provided an equilibration time of less than one hour (in some instances, only about 15-30 minutes) prior to initiation of the toxicity tests. Assuming that the toxicological responses reported by Glover and coworkers were indicative of the kinetics of complexation of Ag by sulfide and/or NOM, it would seem that this would result in the Ag complexation reactions having advanced to only about 10-25% completion by the time the toxicity tests were begun. The complexation of Ag will necessarily continue during the toxicity test, such that the organisms are actually responding to a time variable water chemistry condition. While this confounds the interpretation of the test results, what is clear is that only a limited fraction of the measured CRS would have complexed Ag^+ during the early hours of the test. This is an important point because recent results have shown that aquatic organisms are very sensitive to silver during this initial non-equilibrium period (Kolts et al., 2006; Naddy et al., 2007). The likely explanation is that elevated free metal activity during the initial period of equilibration facilitates transport of the excess metal across the branchial epithelium and into the cell. Then, once the silver is inside the cell, it is no longer in equilibrium with the changing conditions outside the cell, where conditions are improving with respect to the Ag^+ concentration. Further work will be needed to gain an improved understanding of the controlling processes.

As a test of the preceding ideas, an alternative set of predictions was made to see if incomplete complexation of Ag^+ by NOM binding sites and CRS could account for the over-prediction of the observed results by the BLM. The approach was to reduce the complexation capacity (NOM strong binding sites and CRS) to the concentration that was effectively present at the start of the test as a result of the short equilibration time. The calculations were not meant to demonstrate an approach that should routinely be applied when using the BLM in the future. Rather, the calculations were intended to characterize the magnitude of the effect that incomplete complexation of Ag^+ could have had on the Dataset 1 toxicity tests, given the limited equilibration time that was provided. The analysis showed that reasonable results could be achieved if only 20% of the NOM strong binding sites and CRS concentrations were specified as inputs to the BLM (Figure 4-1, filled squares). The LC50 values that were predicted in this case provided reasonable estimates of the natural water LC50 values. The overall range of measured dissolved LC50 values was 0.54-12.1 µg/L (0.90-12.1 µg/L for tests with filtered water samples) and the BLM-predicted LC50 values were generally within a factor of two of the measured values. The use of 20% of the CRS and strong binding sites is qualitatively consistent with an equilibration time of < 1 hour and the rate of complexation that is indicated by the toxicity data of Glover et al. (2005a) (Figure 4-1, inset), of about 20% per hour.

As discussed below, the effect of CRS on the results for Dataset 1 was greater than for Dataset 2 (Section 4.2.2; Bielmyer et al., 2007), even with the incomplete kinetics for Dataset 1. This is because the measured CRS concentrations for Dataset 1 (50 - 211 nM in 13 of the 15 test waters that were analyzed for CRS) tended to be much higher than for Dataset 2 (a maximum CRS of 49 nM). The high CRS levels in these natural water samples of Dataset 1 explains why the average measured LC50 was 8.7-fold higher for these natural waters than it was in the paired hardness adjusted reference waters (i.e., average LC50 values of 5.3 and 0.6 µg/L, respectively).

The problem of insufficient equilibration time precludes the direct use of the natural water results of Dataset 1 for either model calibration or model validation purposes. Even so, these data highlight two important observations that should be relatively independent of effects related to equilibration time. First, in spite of the apparent reduction in the measured site water LC50 values caused by incomplete complexation, the water effect ratios (WER = site water LC50 / hardness adjusted lab water LC50) that resulted were substantial, ranging from >3.0 to >11.6 (average for Desjardins Canal). Further, if the original BLM results are used to estimate what the site water LC50 values would have been with sufficient time provided for equilibration, it is estimated that the WERs would have ranged from 19.5-53.5. While these WERs might appear to be high, they are lower and reasonably consistent than the WER reported by Erickson et al.,

(1998) for a similarly sensitive organism, *D. magna* (lab water LC50 = 0.58 μg/L, versus an average *C. dubia* LC50 of 0.58 μg/L and range of 0.21-1.03 μg/L) .That is, subsequent to a 72-hour equilibration time, Erickson measured a *D. magna* WER for St. Louis River water of 60.3. This high WER was a reflection of the high DOC of the site water (18.3 mg/L) and perhaps an elevated CRS (CRS was not measured). While the reference water was in this case not adjusted to the hardness of the St. Louis River water, this factor was not likely important given the slight effect of hardness on Ag toxicity to *D. magna* (Chapter 3.0).

The second observation to be made on the basis of the Dataset 1 results is that the dissolved Ag LC50 values that were measured with unfiltered and filtered test waters yielded comparable results, while results with lab water that was amended with pre-concentrated NOM to the site water DOC (5 RO samples and 3 RE samples) tended to be more variable. That is, the dissolved LC50 values were in some cases comparable to the raw water results (i.e., results of tests conducted with unfiltered and filtered samples), but in other cases they yielded much lower LC50 values than were obtained when the DOC in the test water had not been pre-concentrated prior to testing (Appendix B, Table B-2). Thus, toxicity testing with NOM concentrates should proceed with caution until methods have been developed that ensure that the nature of the NOM has not been adulterated by the procedures used to concentrate the samples.

4.2.2 Dataset 2- *C. dubia* and *P. promelas* in Natural Waters from North America

The second set of natural water test results, Dataset 2, was generated for the purpose of validating the Ag BLM for two different organisms, *C. dubia* and *P. promelas*. The surface water samples were obtained from eight different locations in the United States and Canada. These samples were analyzed for the water quality variables considered to control the bioavailability and toxicity of Ag (Bielmyer et al., 2007). The sites where these natural water samples were obtained and the analytical results for each are summarized in Appendix B, Table B-3. The samples exhibited a rather wide concentration range for each of the analytes: pH = 6.84 to 8.14, DOC = 1.8 to 12.4 mg/L (150 to 1030 μM), Na^+ = 3.2 to 103 mg/L (139 to 4,480 μM), Cl^- = 3.7 to 155 mg/L (104 to 4,370 μM), Ca^{2+} = 3.2 to 68.8 mg/L (79 to 1,720 μM), Mg^{2+} = 0.85 to 18.5 mg/L (35 to 761 μM) and K^+ = 0.23 to 11.5 mg/L (5.9 to 294 μM). The total hardness varied from 11 to 197 mg/L as $CaCO_3$ with molar ratios of Ca/Mg = 0.87 to 6.88. CRS levels were also variable but tended to be relatively low in these waters (CRS ~1 to 49 nM) in comparison to the waters of Dataset 1. The laboratory reference waters, which were either tap water or 10% tap water plus 90% deionized water, had somewhat atypical characteristics. In particular, the DOC (4.5 mg/L) and Na^+ (1 mM ~ 23 mg/L) of the undiluted tap water were somewhat higher than would normally be expected for tap water.

4.2.2.1 Initial BLM Results

The toxicity test methods used and the toxicity test results for *C. dubia* and *P. promelas* for the second set of eight natural waters (Dataset 2) are described in detail by Bielmyer and co-workers (2007). The test protocol that was followed included a 24-hour equilibration period between the exposure water and the added $AgNO_3$ prior to the addition of the organisms. The originally predicted *C. dubia* LC50 results were reported by Bielmyer and colleagues (Bielmyer et al., 2007). These initial predictions incorporated competition between Ag^+ and other metals for complexation. As described by Bielmyer and coworkers, the BLM predictions were relatively good for *C. dubia* in comparison to the predictions for *P. promelas*. However, three of nine predicted LC50 values were more than a factor of 2 below the measured values (Horsetooth

Reservoir and Trout Lake, both low ionic strength waters, and St. Louis Bay, the water with the highest measured LC50). The results for *P. promelas* were also problematic, with the overall response predicted by the BLM (~7.0-33.9 μg/L) being considerably less than the response exhibited by the data (2.0-44.1 μg/L). Interestingly, and in contrast to the *C. dubia* results, the *P. promelas* LC50 values in the low ionic strength and low hardness waters (Horsetooth Reservoir and Trout Lake, as well as the diluted reference water), were over-predicted by more than a factor of two by the BLM. The LC50 for the University of Miami tap water sample was also over-predicted by nearly four-fold.

4.2.2.2 Exploratory Analyses and Recalibration of the BLM

The outcome of the initial results for Dataset 2 led to the need to complete additional exploratory data analyses and laboratory investigations. These efforts were directed at gaining a better understanding of the factors that might explain the deviations that were evident between the BLM predictions and the data. The goal was to achieve an improved calibration of the BLM by incorporating refinements that could be reasonably justified on the basis of these efforts. Some of the more important data analyses that were completed were related to investigations of the effects of: 1) an extended holding time prior to analysis on the analytical results for selected samples, 2) metal competition on complexation of silver by CRS, and 3) an alternative representation of NOM complexation of Ag. The results of the laboratory investigations that were carried out will be described in the following section (4.2.2.3).

With regard to the first item listed above, consideration was given to the possibility that an extended holding time for three of the samples (St. Louis Bay, Miller Creek and Everglades) prior to the CRS analysis, an inadvertent result of instrumentation problems, may have biased the CRS measurements and adversely affected the agreement between model and data. This could have occurred if there was loss of CRS during the extended holding period, thereby compromising the CRS measurement. The result would be a measured CRS concentration that was lower than what was present in the water during the toxicity test. The BLM predicted LC50 for these samples, therefore, would be based on incorrect water chemistry data and the low CRS value would bias the model results to predict lower LC50 values than were observed. To the degree this problem occurred, it would necessarily affect the predicted results for both *C. dubia* and *P. promelas*, since both predictions relied on the same analytical chemistry data. If the CRS values used by the model were increased by an amount that maintained the *C. dubia* prediction within acceptable limits (i.e., no more than a factor of 2 greater than the measured LC50), the same CRS change would also increase the *P. promelas* predictions to a level that was within a factor of 2 of the measured LC50 as well, for two of the three samples in question. That is, the adjusted *P. promelas* LC50 of 25.6 μg/L for the St. Louis Bay sample was in reasonable agreement with the observed LC50 of 44.1 μg/L. Similarly, the adjusted LC50 of 11.8 μg/L for the Miller Creek sample was in good agreement with the measured LC50 of 16.0 μg/L. However, a similar sensitivity calculation would only have a marginal effect on the prediction for the Everglades sample (measured *P. promelas* LC50 = 44 μg/L), because the BLM was already predicting a *C. dubia* LC50 (1.15 μg/L) that was reasonably close to the measured value (1.45 μg/L). As a result, the allowable increase in CRS would only lead to a marginal improvement in the *P. promelas* LC50. Conversely, increasing the CRS by an amount that would achieve agreement with the fathead minnow LC50 for the Everglades sample would lead to a considerable overestimate of the *C. dubia* LC50 for the same sample. These latter results are in keeping with the results of Glover and coworkers (Glover et al., 2005b) who investigated if a delay in the analysis of CRS subsequent to toxicity tests with *D. magna* might have compromised their results. They concluded that this was not the case. In summary then, while the extended holding time for three of the samples of Dataset 2 may have contributed to the underestimation of the LC50s by the BLM, this explanation was necessarily conjectural and, further, it failed to fully account for the discrepancies that were observed.

The initial BLM results that were reported by Bielmyer et al. (2007) incorporated competition between Ag^+ and other cationic metals for complexation by CRS. This approach had been originally adopted as a possible explanation of the low LC50 results of Dataset 1. That is, complexation of CRS by other metals would reduce the effective concentration of CRS that was available to form a complex with Ag, thereby leading to lower LC50 predictions. However, ongoing analyses indicating that complexation kinetics could readily account for the reduced LC50 values of Dataset 1, as well as a preponderance of information to indicate that Ag would likely out-compete most other metals for CRS (Bell and Kramer, 1999), led to a discontinuation of this approach. Elimination of metal competition resulted in the BLM predictions summarized on Figure 4-2 for *C. dubia* (left panel) and *P. promelas* (right panel). The layout is the same as for Figure 4-1, except that the 95% confidence limits are included for the measured site water LC50 values. As shown on the left, the range of measured *C. dubia* LC50 values in the natural waters of Dataset 2 varies by about 28-fold, from 0.34 µg/L to 9.52 µg/L. This range reflects the wide variation in water quality characteristics of the natural water samples that were included in Dataset 2. Elimination of metal competition resulted in an increase in the predicted LC50 for selected samples, particularly those having relatively high levels of both CRS and competing metals (e.g., the Desjardins Canal and Everglades samples). However, it failed to account for the under-prediction of the LC50 for the three samples that were most problematic (i.e., Horsetooth Reservoir and Trout Lake, both low ionic strength waters with low CRS, and St. Louis Bay, one of the samples with an extended holding time prior to analysis of CRS).

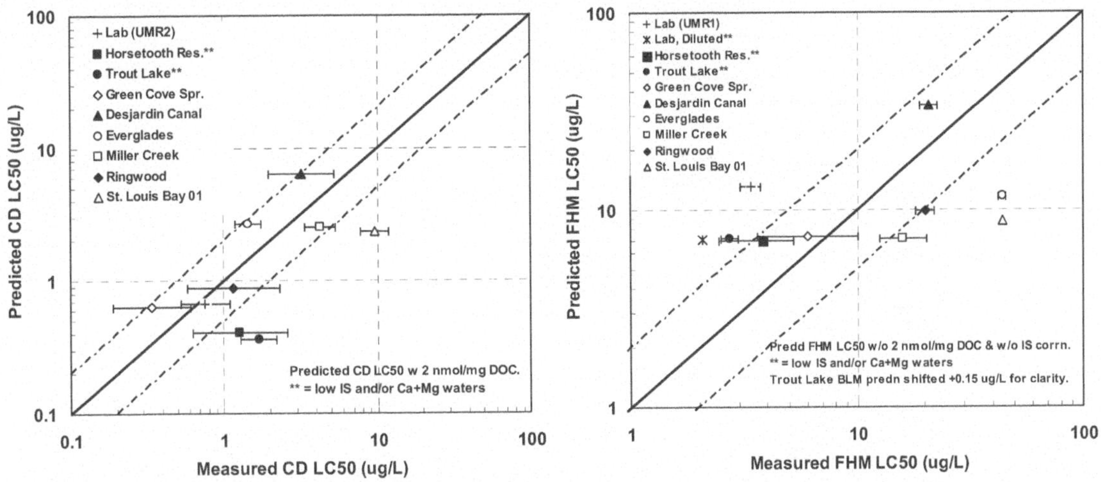

Figure 4-2. Preliminary BLM-Predicted versus Measured LC50 Values for *C. dubia* (left) and *P. promelas* (right) for Natural Water Dataset 2, without Metal Competition. (Data: Bielmyer et al., 2007 and Appendix B, Table B-3)

The elimination of metal competition with Ag^+ for CRS had only a marginal effect on the predicted LC50 results for the *P. promelas* that were exposed to Ag in these same test waters. These results are shown on the right side panel of Figure 4-2. In this case most of the indicated ranges represent the 95% confidence limits of the measured LC50 results. Two exceptions are the ranges shown for two low ionic strength waters (Trout Lake and Horsetooth Reservoir), which indicate the ranges in the LC50 measurements for acclimated and unacclimated LC50 results (discussed in Section 4.2.2.3). The *P. promelas* results of Figure 4-2 are very similar to those of Bielmyer et al. (2007), because they were only marginally affected by the elimination of metal competition. The *P. promelas* results are not nearly as sensitive as the *C. dubia* results to the elimination of metal competition because the effect of Ag complexation by CRS is small in comparison to the relatively high *P. promelas* Ag effect levels in these natural water samples.

Pursuant to the third possibility identified above, consideration was given to modifying the manner by which NOM binding sites for silver were represented in the BLM. The chemistry calibration had been relatively insensitive to the concentration of strong binding sites that were specified (Chapter 2.0) over a range that did influence the fit of the toxicity data. The assignment of 2 nmol Ag^+ binding sites per mg DOC (Log K = 12.0), in association with an adjustment to the LA50, provided a reasonable BLM-predicted LC50 response to DOC (i.e., ~0.2 µg Ag/L per mg DOC/L), while at the same time not adversely affecting the chemistry calibration. The results presented previously, in Chapter 3.0 and on Figure 4-1, were based on this parameterization. In the case of *C. dubia*, the LA50 that is used with the updated version of the chemistry model was evaluated on the basis of lab water test results of (Naddy et al., 2008). The net effect of the change in LA50, in combination with the updated representation of NOM binding, was that there was only a limited change in the predicted LC50 values for the low DOC lab waters. Application of this parameterization of the BLM to Dataset 2 led to the *C. dubia* results of Figure 4-3. As shown, there is a generally improved agreement between model and data, with the predicted LC50 for only one of the site waters (Everglades) deviating by more than a factor of 2 from the observed LC50. The predictions for the three LC50 values that were previously under-predicted by the BLM are considerably improved as a result of this change and they are now within a factor of two of the observations. The improvement reflects the combined effects of the change in the DOC response of the model as well as the adjustment that was made to the LA50 on the basis of the lab water test results of Naddy et al. (2008).

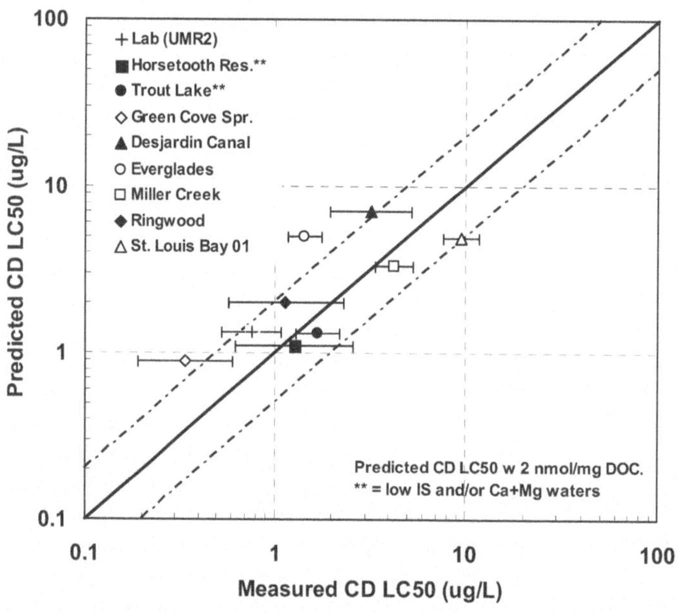

Figure 4-3. Comparison of BLM-Predicted and Measured LC50 Values for *C. dubia*, Natural Water Dataset 2. (Data: Bielmyer et al., 2007 and Appendix B, Table B-3)

Figure 4-4 presents a summary comparison of model and data for the composite dataset for *C. dubia*. Note that the results for the natural water and reconstituted water samples from Dataset 1 are not included in this summary because these test conditions did not conform to near-equilibrium conditions, an implicit assumption of the BLM. The model demonstrates that it has reasonably good explanatory power for the remaining results, for test conditions that were considered to be representative of near-equilibrium conditions. That is, the BLM-predicted LC50 results are in most cases within a factor of 2 of the observations over an LC50 range of about 0.1 to 10 µg/L. Inspection of the results of Figure 4-4 suggests that, as calibrated here, there may be a slight bias on the high side for the predicted results. This in part reflects the approach that was followed, the evaluation of the LA50 on the basis of the lab water data only, and then using this LA50 to predict the LC50 values for Dataset 2. Other approaches that might improve the agreement between model and data would be to further adjust the calibration by using the complete *C. dubia* dataset (including Dataset 2) to evaluate the LA50 or, alternatively, to simply reduce the LA50 slightly.

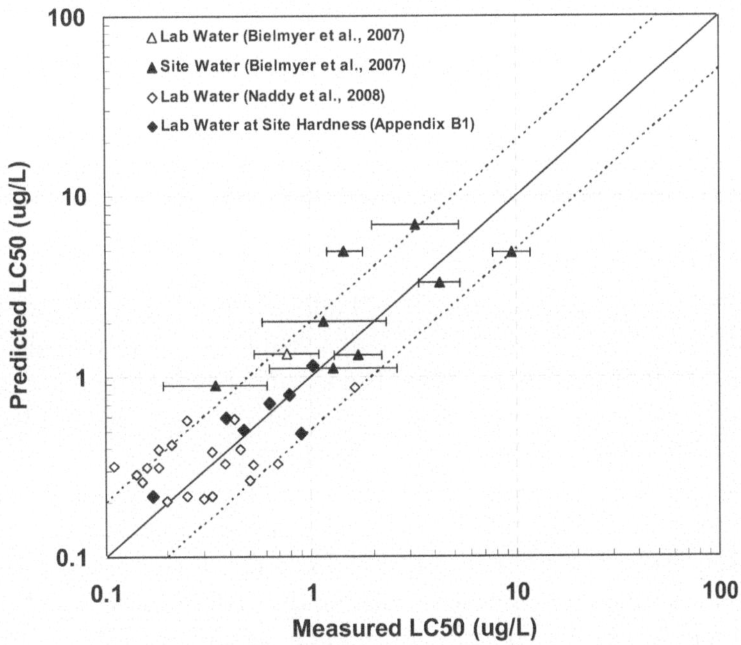

Figure 4-4. Comparison of BLM-Predicted and Measured LC50 Values for *C. dubia*, Composite Dataset. (Data: Bielmyer et al., 2007; Naddy et al., 2008 and Appendix B, Tables B-1 and B-2)

The preceding adjustment to the chemistry model must necessarily be applied to all organisms. However, the significance of this change on the predicted LC50 will vary with organism sensitivity. The reason is that the change in LC50 is only about 0.2 µg/L of Ag per mg/L of DOC. A difference of this amount will be relatively important for organisms that are sensitive to silver such as *C. dubia* (LC50 of <1 µg/L in lab water). However, the effect will diminish as sensitivity decreases, with its importance for an organism such as *P. promelas* (effect levels of 2 µg/L or greater) being considerably less than the effect on *C. dubia*. It is because of this that the previously described change related to NOM complexation has only a limited effect on the *P. promelas* predictions for Dataset 2 (compare right panel of Figure 4-2, to the updated results on Figure 4-5). It was clear from these results that additional factors were operative and that the *P. promelas* dataset required still further consideration.

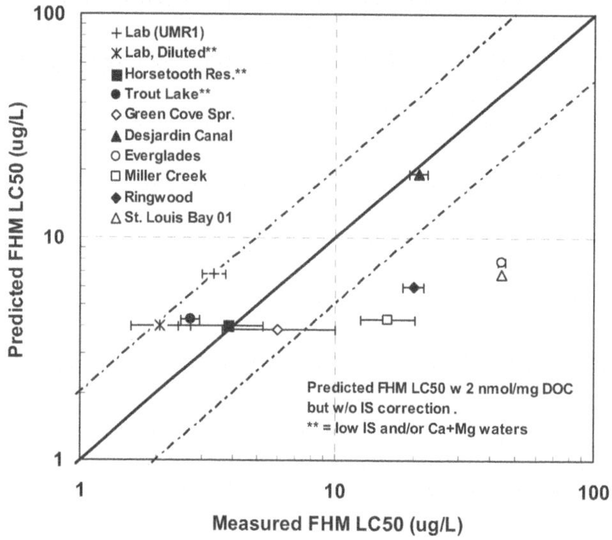

Figure 4-5. Comparison of BLM-Predicted and Measured LC50 Values for *P. promelas*, Natural Water Dataset 2, with 2 nmol/mg DOC but without an IS Adjustment of the LA50. (Data: Bielmyer et al., 2007 and Appendix B, Table B-3)

4.2.2.3 Physiological Investigations with *P. promelas*

Prior to the WERF investigations the BLM had been applied to toxicity test results for 4-day, 28-day and 42-day old *P. promelas* that were exposed to Ag in systematically modified lab waters (Erickson et al., 1998; Karen et al., 1999; Bury et al., 1999a). Pursuant to the WERF investigations, most of the *P. promelas* toxicity tests made use of <1-day old fish that were grown in University of Miami dechlorinated tap water for 7 days prior to being exposed to Ag in the natural water samples. The ongoing review of the natural water test data as well as the earlier test results in lab waters indicated that the *P. promelas* that were exposed to Ag in low IS and/or low hardness waters tended to be relatively sensitive in comparison to *P. promelas* that were exposed to Ag in higher hardness or IS waters. That is, the difference in Ag effect levels was greater than could readily be accounted for by differences in the chemistry of the exposure waters. A similar result was recently reported for *P. promelas* exposed to Cu, where BLM predictions also tended to overestimate the Cu LC50 values that were measured in soft, low IS water (Van Genderen et al., 2005).

The two natural water samples of Dataset 2 that were lowest in IS and hardness were the samples from Trout Lake, Ontario and Horsetooth Reservoir, Colorado (total hardness = 11 and 34 mg/L as $CaCO_3$, respectively). The dechlorinated reference water sample that was diluted 10-fold with deionized water was also low in hardness and IS. The BLM results shown on Figure 4-5 include predicted LC50 values for these samples that are within a factor of two of the measurements, but the use of a constant LA50 necessarily leads to a marked under-prediction of the LC50 results in four of the other natural waters (Miller Creek, Ringwood, Everglades and St. Louis Bay). Alternatively, increasing the LA50 to improve the agreement between model and data for the LC50 values in these four samples would lead to an overestimation of the LC50 at the low end of the measured range of LC50 results, including the LC50 values for fish that were exposed to silver in low IS water. These calculations suggest that the fish exposed to silver in the low ionic strength water are relatively sensitive to silver, at least more so than can be attributed by the BLM to differences in water chemistry.

One hypothesis that was proposed in an effort to explain the preceding results for the low IS waters was that the fish that were tested in these waters needed additional time for acclimation. The toxicity tests in the low ionic strength waters were repeated, but in this case the fish were provided with a 7-day period of acclimation to the exposure water prior to the toxicity test (rather than acclimation in University of Miami tap water) (Bielmyer et al., 2007). Counter to expectations, the results showed that the acclimated and unacclimated LC50 values were *not* significantly different (based on overlap of 95% confidence limits):

	Unacclimated LC50 (μg/L)	**Acclimated LC50 (μg/L)**
Trout Lake, Ontario, Canada	2.49 (1.88 – 3.30)	2.96 (2.41 – 3.64)
Horsetooth Reservoir, CO	5.23 (4.0 – 6.23)	2.44 (1.21 – 4.89)
10% UM reference water	1.99 (1.70 – 2.34)	2.18 (1.69 – 2.81)

Note: LC50 (+/- 95% Confidence Limits)

The acclimated and unacclimated LC50 results for these two natural water samples were indicated by the ranges that were shown previously for these samples on Figure 4-2 (right panel; the ranges for the other natural waters and for the acclimated test in UM reference water correspond to the 95% CL for the measured LC50 values). The insensitivity of the *P. promelas* Ag results to use of low or high hardness acclimation waters is consistent with results reported by Erickson and coworkers in tests with 28-day *P. promelas* that were exposed to copper (Erickson et al., 1997). Their results showed that a 12- to 14-day period of acclimation to either low or high hardness water (0.91 or 4.2 mEq/L CaC12) resulted in little if any indication of acclimation effects on subsequent 48-hour Cu EC50s for 28-day *P. promelas*.

Additional tests were performed in an effort to obtain an improved understanding of the preceding results for unacclimated and acclimated test conditions. The objective was to determine if the ionic strength of the exposure water was leading to a change in the kinetics of branchial Ag accumulation (i.e., Ag uptake at the gill, the site of action of acute Ag toxicity to fish) by *P. promelas* (Bielmyer et al., 2008). These tests were conducted by acclimating one set of fish to a low ionic strength soft water and a second set to a higher ionic strength moderately hard tap water. Both sets of fish (weight = 2.26 ± 1.0 mg) were then exposed to 0.2 to 12 μg/L Ag, in the same soft water, for a period of 3 hours. It was found that fish acclimated to the low ionic strength water accumulated Ag at a faster rate and to a higher level than fish acclimated to the moderately hard water (Figure 4-6), even though they were exposed to Ag under identical conditions. A possible explanation of these results is that fish acclimated to low ionic strength

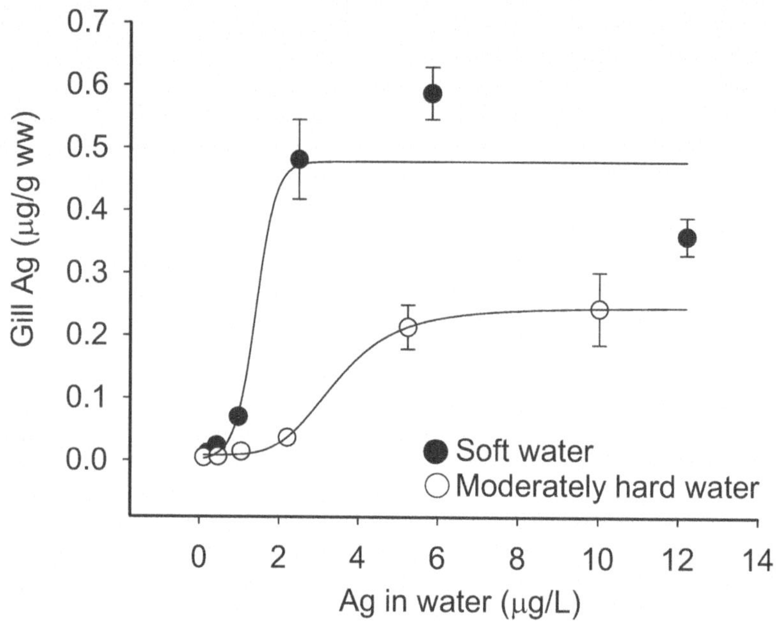

Figure 4-6. Effect of Acclimation Water Quality on 3-hour Gill Ag Uptake in Soft Water by *P. promelas*. (Data: Bielmyer et al., 2008)

soft water (low in Na^+ and other ions) responded to these conditions by up-regulating their ability to take up Na^+ (and other ions) as they acclimated to the water. An adaptation of this nature would facilitate maintenance of ionoregulatory homeostasis and the ability of the fish to survive under such conditions. However, this same physiological change would also be expected to lead to an inadvertent increase in Ag^+ uptake because Ag^+ shares some of the same ion uptake pathways as Na^+ (i.e., it inhibits Na^+ influx and mimics Na^+ in the entry to the cell via an apical-located H^+-coupled Na^+ channel; Bury and Wood, 1999; Bury et al., 1999b). The increased rate of Ag^+ uptake that was observed by the fish that were acclimated to the low IS water would be expected to offset, to some degree at least, the ionoregulatory benefit that would otherwise accrue from this same period of acclimation. This line of reasoning is consistent with the toxicity test results that showed that prior acclimation to the low IS test water did not result in a significant reduction in the sensitivity of *P. promelas* to Ag. While considered together these test results provide further insight into the effect of acclimation to low ionic strength waters on Ag uptake, they do not explain why the fish that were exposed to Ag in the low ionic strength natural waters (acclimated or not) seemed to be more sensitive to Ag (i.e., the LC50 values were relatively low) than could be accounted for by the effect of water chemistry on the LC50 that was predicted by the BLM.

4.2.2.4 Modification of the BLM for Ionic Strength Effects

It was clear from these and earlier lab water test results that a modification was needed to improve the ability of the BLM to predict *P. promelas* effect levels over the full range of the water quality characteristics of the natural water samples. Exploratory data analyses, including an analysis of residuals (relationships between deviations of BLM predictions from the data with other water quality characteristics), did not lead to the identification of any specific water quality characteristics that could account for the observed test results. It was recognized, however, that the calculated BL:Ag accumulation level, evaluated at the measured Ag LC50, tended to decrease with decreasing ionic strength, at least for the *P. promelas* that were more than 4 days old (Wu et al., 2008). Because it was previously assumed that this lethal accumulation level was a constant for any set of water quality characteristics, a simple approach to account for the deviation of the model from the data was to specify the LA50 as a function of ionic strength (Wu et al., 2008). This was done such that *P. promelas* in a given age class had a constant age-specific LA50 for IS > 0.01 M (see Chapter 3.0 and Wu et al., 2008) and an LA50 that decreased with IS for IS < 0.01 M (one slope) for all age classes > 4 days. It should be clear that this approach is not mechanistically-based and that it might be more correctly represented by some as a yet to be understood ion-specific response that occurs in low ionic strength water. However, adjustment of the LA50 in this manner does provide an empirical correction that accounts for the LC50 response of older, larger *P. promelas* that are exposed to Ag in low IS waters. Implementation of this approach improved the agreement between the predicted LC50 values and the data for the natural water test results for *P. promelas* (Figure 4-7, filled circles), with the LC50 for the undiluted reference water that had a DOC of 4.5 mg/L (i.e., the University of Miami tap water sample) being the only *P. promelas* data point of Dataset 2 that is still over-predicted. The predicted LC50 values for the Miller Creek, St. Louis Bay and Everglades samples continue to be somewhat low, about a factor of 2 or more below the measured LC50 values (16.0, 44.1 and 44.0 μg/L, respectively). While not providing a complete explanation, the extended holding time and potential effects of this on the measured CRS concentrations for these samples remains as a factor that may be contributing to these deviations. The *P. promelas* BLM results for other age classes of *P. promelas* (4-, 28-, and 42-day old fish), are also shown for purposes of comparison and are seen to be in generally good agreement with the *P. promelas* data that were obtained in tests with systematically modified lab waters.

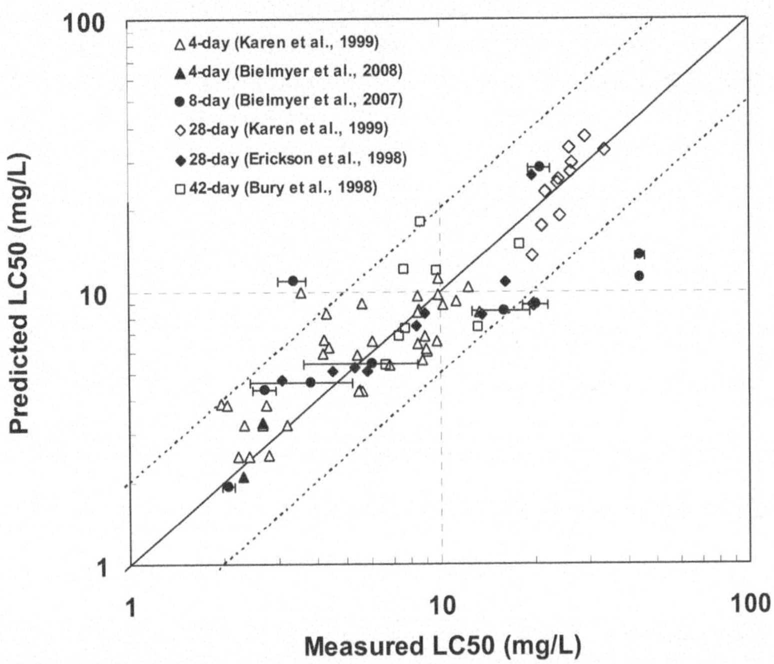

Figure 4-7. Predicted LC50 versus Measured LC50 for *P. promelas* Exposed to Ag in Natural Waters and in Laboratory Waters (with IS adjustment except 4-day old fish).

4.3 Summary

While the natural water test results presented a challenge for the WERF BLM development efforts, they also served as a stimulus for investigations that have led to an improved understanding of how physiology affects the sensitivity of aquatic organisms to Ag. These investigations have also led to what is considered to be an improved version of the Ag BLM itself. The *C. dubia* natural water results for Dataset 2 demonstrate that the current implementation of the BLM is in most cases able to simulate the LC50 for this Ag-sensitive organism within a factor of 2 over a relatively wide range of water quality characteristics. The results of Dataset 1 were problematic in that the test conditions were not entirely consistent with the underlying requirement of the BLM, that chemical equilibrium conditions apply. Even so, these test results highlighted the utility of the BLM as a diagnostic tool which can be used to better understand experimental results. The Dataset 1 test results provided a conservative demonstration of the significant mitigating effect that water quality can have on Ag bioavailability and effect levels. It was possible to make use of the BLM to evaluate what the mitigation potential would likely have been had a longer equilibration time been used.

The effects of fish age (or size) and ionic strength were shown to markedly influence the sensitivity of *P. promelas* to silver. This led to the need to employ empirical adjustments to the LA50 as a way to account for these factors. It is acknowledged that it would be desirable to have a more fundamental understanding of the reasons for these responses. However, it is significant to note that from an implementation perspective, the most sensitive age class of *P. promelas* that was considered, 4-day old fish (triangles on Figure 4-7), did not exhibit a similar systematic variation in effect levels with IS, even at an IS as low as 0.0048 M. Thus, as long as this most sensitive age class is used for evaluation purposes (e.g., ecological risk assessment or WQC derivation), then gaining an improved understanding of the mechanism of ionic strength effects and the development of a better way to represent age/size and IS effects in the BLM are endeavors that are more of academic interest than an immediate practical need. At the same time, longer term advances in understanding related to these areas should ultimately lead to an improved version of the BLM, both for Ag as well as for other metals, generally.

CHAPTER 5.0

DISCUSSION, CONCLUSIONS, AND RECOMMENDATIONS

5.1 Introduction

As is often the case for any research effort, the WERF Ag BLM program answered many questions and, at the same time, identified others that needed to be addressed. Important technical areas that project-related investigations focused on were the chemistry of Ag, sulfide and NOM interactions, the effect of organism size and/or age on sensitivity to Ag, the effect of ionic strength on sensitivity to Ag, and the effect of Cl^- on the bioavailability and toxicity of Ag to aquatic organisms. Not surprisingly, some of the issues related to these topics have yet to be fully resolved. Overall, however, the Ag BLM investigations have led to an improved understanding of the bioavailability and toxicity of Ag in aquatic environments. Equally important, it has also added to the database that is available for use in the derivation of a BLM-based acute WQC for Ag. With regard to the Ag BLM itself, the result is an improved capability to predict site-specific effect levels for Ag in aquatic environments and a more rigorously tested and broadly applicable version of the acute BLM for Ag. This concluding chapter includes a discussion of some of the important technical areas that were addressed during this project and it summarizes the progress that has been made toward the achievement of the longer term management objectives that were outlined in the Introduction (Chapter 1.0). Recommendations for study that should lead to further advances in the current understanding of the bioavailability and effects of Ag and other metals in aquatic environments are also provided.

5.2 Discussion

The Ag BLM investigations were designed to test the version of the BLM that was available at the initiation of this project, to satisfy previously identified chemistry and toxicity data gaps, and to use the results of the laboratory investigations to further refine and strengthen the technical validity of the Ag BLM. While much was learned from the Ag BLM investigations that were carried out to address these needs, these same studies also elicited additional questions that needed to be addressed. This led to investigations that were carried out to gain a better understanding of how chemistry and physiology affect the bioavailability and toxicity of Ag and how to best represent these factors in the BLM. With regard to the aquatic chemistry of Ag, interactions with sulfide and with NOM-related strong binding sites were particularly important, given the demonstrated ability of complexation to mitigate the effects that might otherwise result from the typically low exposure and effect levels of Ag in aquatic environments. The importance of complexation kinetics was also highlighted. Other studies took advantage of recent advances in organism physiology as a way to understand the effects of organism age (or size) and ionic strength on sensitivity to Ag. With regard to assessing the protective effect of chloride on Ag toxicity it was necessary to consider both chemistry and physiology to gain an improved understanding of the test results. Highlights of each of these technical areas are discussed below. While it will be evident from the results that the chemical and physiological interactions that affect the bioavailability and toxicity of Ag are complex and the toxicological responses are at times subtle, an important benefit of the WERF investigations is that an improved understanding of the bioavailability and toxicity of Ag has been realized.

5.2.1 Interactions of Ag with Sulfide and NOM-binding Sites

Silver toxicity is mitigated by the formation of complexes with various inorganic and organic ligands. Two of the more important such ligands are NOM and sulfide, with the latter often found to be associated with NOM. As discussed below, the situation is complicated by the observation that the rate of complexation of Ag by these relatively complex ligands is not necessarily as fast as is complexation by inorganic ligands such as Cl^-.

5.2.1.1 Complexation of Ag by NOM and Sulfide

Sulfide is an important ligand that forms strong complexes with many metals, thereby reducing their bioavailability to aquatic organisms. It is a particularly important ligand for Ag because the typically low environmental levels of Ag are often comparable in magnitude to the complexation capacity of environmentally relevant sulfide levels. Complexation of Ag by inorganic sulfide will therefore mitigate the adverse consequences that might otherwise result at the low levels of Ag that have been shown to adversely affect some aquatic organisms in laboratory waters. Application of the BLM to the POP experiments (Chapter 3.0) showed that the BLM was able to simulate the protective response that had been demonstrated experimentally for the sensitive invertebrate, *D. magna*, and for a somewhat less sensitive fish, *O. mykiss*.

Another important area that was elucidated by the experimental program and related modeling analyses was the complexation of Ag by organic matter, including sulfide that is associated with organic matter. It was found that these reactions do not necessarily occur over a time scale of minutes, as might be the case for the formation of inorganic Ag sulfide complexes. Instead, several hours or longer may be required to approach equilibrium conditions. The hindered rate of complexation is thought to be a result of the rate of transfer of Ag into the large macromolecules that NOM is comprised of. As discussed below, it was concluded that a slow rate of Ag complexation by CRS and NOM was the cause of the discrepancy between the equilibrium-based BLM LC50 predictions and the markedly lower LC50 measurements of Dataset 1. This finding is consistent with results of other parallel investigations (i.e., contemporary but non-WERF related investigations) that have shown that 3 hours or longer may be required for Ag to approach equilibrium conditions (Glover et al., 2005a; Kolts et al., 2006; Naddy et al., 2007). Other investigations have provided similar results for Cu interactions with organic matter (Ma et al., 1999, for *C. dubia*; Santore et al., 2001 and U.S. EPA, 2002b, for *P. promelas*).

It is not essential that sulfide be routinely measured in all situations. For example, if the dissolved Ag is expected to be at levels that are well below adverse effect levels in lab waters, then the potential for effects will be low regardless of the sulfide concentration that is present. At higher levels of dissolved Ag other characteristics of a water body, including pH, DOC, Na^+ and Cl^- may also alter the magnitude of the expected acute effect levels such that this information may be sufficient to rule out the likelihood of effects and the need to measure sulfide. When this is not the case, then the analysis of sulfide may be needed to fully assess the potential for adverse effects due to Ag. Similarly, the evaluation of a BLM-based WQC may or may not be sensitive to the consideration of sulfide levels. If site-specific sulfide levels are known to be very low, then it will be of little consequence if sulfide is neglected. Alternatively, if sulfide levels tend to be high, then evaluation of a BLM-based WQC for Ag will be conservatively low because it would neglect the potential for sulfide to mitigate toxicity due to Ag. Although CRS has not routinely been measured by most analytical laboratories, updated analytical methods are available for use (Bowles et al., 2003; Kramer et al., 2007).

The effect of sulfide on the chronic toxicity of Ag to aquatic organisms has not been well studied to date. However, the potential for sulfide to mitigate the chronic effects of Ag on *C. dubia* has been demonstrated (Naddy et al., 2007). As a result, there is likely to be added utility to the measurement of CRS for the purpose of an effects assessment that considers chronic effects due to Ag.

5.2.1.2 Complexation Kinetics

Inconsistencies between predicted and measured LC50 results were most evident for Dataset 1. These toxicity tests exposed *C. dubia* to silver in a variety of natural water and wastewater effluent samples and in lab water that was amended with NOM that had been concentrated from these same site water samples. Tests were also carried out in lab water that was adjusted to the hardness of the site waters. The discrepancies between model and data that were evident for the samples with elevated NOM were attributed to the low LC50 values that resulted from an experimental artifact. That is, the measured LC50 values were lower than would have otherwise been the case due to the provision of less than one hour for equilibration of the added $AgNO_3$ with the CRS and NOM binding sites prior to the addition of the test organisms. This prevented the Ag complexation reactions from reaching near-equilibrium conditions and resulted in the partial manifestation of the mitigating effect of these interactions on silver bioavailability and toxicity for this dataset. It is because of this that only a subset of the LC50 values for Dataset 1, the test results obtained with low organic matter lab water adjusted to site water hardness, were considered to be appropriate for use as a direct test of the predictive ability of the Ag BLM. The remaining test results, including results obtained with site waters and with lab water that was amended with reconstituted NOM to the original DOC of the site waters, were still of use in that they provided a conservative demonstration of the ability of natural waters and wastewaters to mitigate Ag toxicity. They were also used to compare different techniques for isolating and concentrating NOM.

It is expected that the pre-equilibration of Ag with the test water, prior to addition of the test organisms, is of limited importance for toxicity tests performed with lab water that is low in NOM. The reason is that the kinetics of inorganic speciation reactions are rapid, occurring over a time scale of seconds to minutes (Adams and Kramer, 1998). This expectation is in keeping with the results of Erickson et al. (1998) where it was shown that increasing the time for aging of a low DOC test solution from less than 2 hours pre-equilibration time to 72 hours did not alter the toxicity of Ag to either *P. promelas* (aging was associated with a qualitative decrease in LC50) or *D. magna* (no significant difference in LC50 values). The situation is more complicated in the presence of elevated levels of NOM, since pre-equilibration time and complexation kinetics have been shown to affect the results of toxicity tests with metals (Ma et al., 1999; Glover et al., 2005a). It was for this reason that the Dataset 1 results with waters having elevated levels of NOM and CRS were not used in the calibration of the BLM or as a way to assess its predictive capability. As outlined below, several lines of evidence support this conclusion.

A review of relevant investigations of the effect of pre-equilibration time on the toxicity of metals to aquatic organisms supports the conclusion that the *C. dubia* toxicity tests with natural water samples and with lab water amended with reconstituted NOM (a subset of the results from Dataset 1), yielded low LC50 values as a result of a methodological artifact, the use of an equilibration time of less than one hour. This time was not sufficient for complexation reactions to reach near-equilibrium conditions (an implicit assumption of the BLM) and likely resulted in elevated free Ag concentrations during the early portion of the test. Supplemental WERF investigations with *C. dubia* (comparing results with equilibration times of 5 minutes and of 3 hours and 40 minutes), in combination with the results of Glover et al. (2005a) with *D. magna* (comparing equilibration times of 3 hours and of 24 hours), indicated that the overall effect of the short equilibration time could reasonably be expected to reduce the measured Ag LC50 values by about a factor of 4.6. This is reasonably consistent with the magnitude of the over-prediction of these results with the BLM (Figure 4-1). The time course data for the *D. magna* experiments of Glover and coworkers (Chris Glover, personal communication) and supplemental WERF project data with *C. dubia* both indicated that there was a relatively high rate of mortality early in the test. These observations, considered to be symptomatic of relatively high Ag^+ levels being present during the early hours of the test, provide further support for the conclusion that non-equilibrium conditions prevailed during the Dateset 1 tests with elevated NOM and CRS.

If the discrepancy between BLM predictions and the observations was in fact related to the kinetics of Ag complexation by NOM or NOM-associated CRS, then the expectation would be that the deviation between model and data should increase as DOC concentration increases. This is what was observed. The regression of predicted minus measured LC50 (Y, µg/L) versus DOC (mg/L) yielded Y = 1.4 x DOC ($r^2 = 0.70$). A relationship of this nature was not evident for the balance of the data considered, including the *D. magna* dataset of Karen et al. (1999), discussed below. This provides further support for the proposed explanation of the Dataset 1 results and, further, that this same factor was not seriously impacting the results of the other datasets used for BLM development purposes.

For the reasons discussed above and in contrast to the natural water results, the subset of *C. dubia* studies of Dataset 1 that were performed with lab water adjusted to site water hardness levels were *not* likely to have been materially affected by slow complexation kinetics. This conclusion is in keeping with the excellent ability of the BLM to predict the LC50 values for these tests. Use of these data in combination with the Dataset 1 natural water results (understood to be low for reasons noted previously) to evaluate water effect ratios (WER = $LC50_{site}/LC50_{lab}$) will therefore provide a conservative demonstration of the potential for natural waters to mitigate the toxicity of Ag. That is, site water LC50 values were 3.0-fold to 11.6-fold higher than the corresponding LC50 results obtained with the lab waters adjusted to site water hardness (i.e., the WERs ranged from 3.0 to 11.6). Pursuant to equilibrium-based BLM predictions, it would be reasonable to expect that these WERs would have been as high as 19.3 to 53.5 had test conditions provided a better approximation of near-equilibrium conditions.

The *C. dubia* and *P. promelas* tests of WERF Dataset 2 incorporated a 24-hour equilibration time and, as such, this approach should have prevented kinetic limitations from having a material effect on these results. The rate constant for the complexation reaction (approximately 20% per hour) that was inferred from the results of Glover et al. (2005a) support this expectation. Equilibration times for the other non-WERF datasets that have been used to support the development of the Ag BLM have also been reviewed in further detail. With the limited exceptions noted below, it generally appears to be the case that short equilibration times and incomplete complexation of added silver has not markedly impacted the datasets that have been used.

The *D. magna* dataset of Karen et al. (1999) is perhaps the most important study where detailed information on equilibration time has not been obtained to date and where there remains a question about the potential for these results to have been impacted by a short equilibration time and slow reaction kinetics. The *D. magna* model was calibrated to this dataset, which included a considerable number of test results that were obtained with low organic matter lab water. Related simulations of the results for DOC-amended lab waters worked reasonably well, without any indication of an over-prediction by the BLM, as would be expected to be associated with slow complexation kinetics. Further, there was no indication of a trend of increasing residuals (predicted LC50 – measured LC50) with increasing DOC concentration. Pursuant to recent inquiries, it is possible that the equilibration time was relatively short for these tests (Steve Klaine, Clemson University's aquatic toxicology lab director). Assuming that it was not of sufficient duration to allow for the establishment of near equilibrium conditions, then the measured LC50 results in NOM-amended waters would be expected to be biased on the low side. Given the general consistency between low NOM lab water results and NOM-amended lab water results with the BLM predictions, it appears that this was not the case.

It is unclear why the BLM did not over-predict the LC50 data for the NOM amended subset of this *D. magna* dataset, as it did for Dataset 1. One possibility is that the equilibration time was longer than has been recently recalled. Otherwise, it is not possible to definitively account for the apparent difference in results for these two datasets, where one seems consistent with the BLM predictions and the other one does not. However, the difference in NOM sources (AHA versus NOM) and/or the elevated levels of CRS in the waters of Dataset 1 are characteristics that differentiate these two datasets from each other. Perhaps one or both of these factors exacerbated the well-demonstrated effect of complexation kinetics on the bioavailability and toxicity of Ag for Dataset 1. To the degree the *D. magna* data were impacted by kinetics, the LC50 values would tend to be low and this would, if anything, result in the calibration of the Ag BLM to data that reflect a conservatively low bias. Further, the BLM has been calibrated to a relatively extensive composite dataset which includes a variety of organisms that were tested over a considerable range of water types and water quality characteristics. As such, the overall influence of any one dataset on the BLM calibration is necessarily constrained.

In summary, it is concluded on the basis of the overall weight of evidence, including but not limited to the results of Glover and coworkers (Glover et al., 2005a), time course mortality data for Dataset 1, and an analysis of residuals, that the difficulty in predicting the *C. dubia* results of Dataset 1 was related to the short equilibration time that was used for those tests. As a result, it was decided that this subset of the natural water test results should not be used for calibration of the BLM or for assessing its predictive ability.

5.2.2 Effect of Size/Age on Sensitivity to Silver

As discussed in Chapter 3.0, the analysis of several *P. promelas* datasets in which fish of different sizes (or ages) were tested (Erickson et al., 1998 and Karen et al., 1999 with 28-d fish; Bury et al., 1999a, with > 42 day old fish; Bielmyer et al., 2007 with 8-d old fish) indicated a need to adjust the model to account for size-related differences in sensitivity to Ag. This response is understood to be related to the inverse relationship of Na^+ turnover rate and size (Figure 3-4; Bielmyer et al., 2007; Grosell et al., 2002), resulting in a relatively rapid loss of Na^+ by relatively small fish when the active uptake of Na^+ is inhibited by exposure to Ag.

Approaches for modeling the effect of Na^+ turnover rate on species sensitivity have been proposed (Grosell et al., 2002; Paquin et al., 2002b). However, the current version of the acute

Ag BLM does not explicitly represent Na^+ turnover rate and the incorporation of this feature into the BLM, for a wide range of organism types, would require a considerable model development and testing effort. Alternatively, it has also been suggested that the BLM binding constant could be increased as a way to simulate an increase in species sensitivity (DeSchamphelaere and Janssen, 2002; Bianchini and Wood, 2003). However, if the underlying mechanism of toxicity is the same, involving inhibition of one or more enzyme systems that are common across species (see Grosell et al., 2002; Bianchini et al., 2002b), then it would seem that the binding constant for Ag^+ at the site of action of toxicity to these enzyme systems should be consistent across species. This same argument, that the controlling enzyme systems should be the same and that the biochemical characterization of the interaction between Ag and these enzyme systems should be consistent, is even more likely to be so for individuals of the same species which differ only in size. If the BLM incorporated the capability to perform an ion balance, then the weight normalized binding site density could be *increased* to simulate increasing sensitivity, based on the premise that BL binding site density varies directly with both weight-normalized enzyme activity and weight normalized Na^+ uptake rate (see Paquin et al., 2002b for a more detailed discussion). Absent this ion balance capability, a more expedient approach of adjusting for species or size sensitivity has been implemented, one in which the LA50 is *decreased* to effectuate an increase in species (or age) sensitivity. The question of how best to represent the effect of size or age for the purpose of evaluating a WQC may be more of an academic interest, provided that the most sensitive life stage or age class (4-day old fish, in the case of *P. promelas*) is used to represent a particular species in the species sensitivity distribution that is used to derive the WQC.

5.2.3 Effect of Ionic Strength on Sensitivity to Silver

Another important issue to address is whether or not the low ionic strength modification that was employed for *P. promelas* that were more than 4 days old is also required for other species of aquatic organisms. The reason this might be important is that all organisms must, to some degree, regulate the ionic composition of their internal fluids. As a result it would be reasonable to expect that any organism will, at some level of ionic strength, be adversely affected by such conditions. The following points are relevant to consider in this regard. First, the problem that was encountered with predictions of the 8-day old *P. promelas* test results was not evident for the *C. dubia* results obtained with the same set of natural water samples. Further, the LA50 values that were calculated with the BLM for all of the *C. dubia* LC50 tests did not exhibit a downward trend with decreasing IS (as was the case for ≥8-day old *P. promelas*). Similarly, the LA50 values were also examined for each of the other species that the BLM was applied to and these results also failed to demonstrate the pattern of decreasing LA50 with decreasing ionic strength that was observed for *P. promelas* and judged to be symptomatic of adverse effects due to low ionic strength water.

Laboratory experiments in which *P. promelas* was acclimated for 7 days to a relatively low ionic strength soft water (in the absence of Ag) failed to alleviate the problem. That is, counter to what might be expected, acclimation to the low ionic strength test water rather than acclimation in a moderately hard water did not lead to a significant increase in the LC50 in the low IS water. This was possibly due to a presumed up-regulation of Na^+ transport kinetics in the low ionic strength water, a change that may have inadvertently led to the increase in Ag^+ uptake that was observed (Bielmyer et al., 2008). It is relevant to recognize in regard to these results that the experimental design did not address the potential for effects, positive or negative, that might be associated with acclimation to low levels of waterborne Ag (i.e., Ag was not added to the low ionic strength acclimation waters).

The observation that the 4-day old *P. promelas* did not appear to be adversely affected by low IS water, while older individuals were, raises the question of whether or not an extended period of acclimation to soft water might have altered the *C. dubia* test results and corresponding model predictions. While this was not investigated as part of these studies, it is known that when *C. dubia* are cultured in the low ionic strength Horsetooth Reservoir water the fecundity of the organisms decreases. Thus, the hardness (and alkalinity) of the culture water must be increased to a moderate level to achieve normal rearing conditions (R. Naddy, personal communication). This may indicate that *C. dubia* may not be well suited for long-term existence in very soft waters, such as the unmodified Horsetooth Reservoir water (in the lab or in the field), even in the absence of Ag. In addition, the effect of culturing them in soft water on their sensitivity to Ag is also not well understood. Given these considerations, it would be of use to know if a fish species such as *P. promelas* could subsist for the long term, even in the absence of Ag, in low ionic strength natural waters that would compromise the organism's resistance to Ag under laboratory test conditions. If not, then this would seem to limit the ecological relevance of test results that are obtained in such waters. Related to this matter, information is currently lacking on possible multi-generational effects of low ionic strength water on species sensitivity to Ag.

Finally, as discussed above, the sensitivity of *P. promelas* to Ag decreases with increasing age, with the 4-day old fish (Karen et al., 1999 and Bielmyer et al., 2008) the most sensitive age class for this species. There did not appear to be an effect of ionic strength on the 4-d *P. promelas* Ag LC50 values, even though the test protocols included ~3 days of acclimation to the test waters (at each of three different hardness levels in the case of Karen et al., 1999) prior to exposure to Ag. Because the 4-day old *P. promelas* were more sensitive than the older fish, fish that seemed to be adversely affected by low IS water, it would seem reasonable to use this age class, without an IS adjustment, to derive an acute WQC for silver. Doing so precludes the need to explicitly represent what is acknowledged to be the poorly understood effect of ionic strength (or, of an ion-specific response) on species sensitivity for WQC derivation purposes. This approach is in keeping with the intent of the Guidelines which recommend, in brief, that relatively insensitive age classes not be used in the derivation of WQC (Stephan et al., 1985).

5.2.4 Effect of Cl^- on Silver Toxicity

As discussed in Chapter 1.0, a commonly held view is that the toxicity of a metal is directly related to the free ion activity of the metal (e.g., Ag^+ or Cu^{2+}) and that reaction of the free ion form with other constituents present in the water (i.e., complexation of the free ion by ligands, the formation of more complex forms of the metal) will reduce the availability of the dissolved metal for uptake by an organism (i.e., the FIAM; see Campbell et al., 1995, for a comprehensive review). It would therefore seem reasonable to assume that increasing the level of Cl^-, which readily forms a number of chloro-complexes with Ag, would reduce the toxicity of dissolved Ag. While this was clearly the case when *O. mykiss* and *C. dubia* were exposed to Ag (Bury et al., 1999a; Naddy et al., 2008), other tests have failed to demonstrate that a significant protective benefit was provided by elevated Cl^- levels (e.g., Erickson et al., 1998 and Bury et al., 1999a for *P. promelas*; and Bielmyer et al., 2008, for *D. rerio*, *F. heteroclitus* and *P. promelas*). Thus, the notion that increasing the Cl^- concentration will mitigate the toxicity of Ag to aquatic life may not be as broadly applicable as might be expected on the basis of the FIAM. Is it possible that chloride protection was simply an artifact of the experimental design in tests with aquatic species where it was observed, or where it was not observed? Could Cl^- protection be a species-specific response? Are there other factors that need to be considered? Is it appropriate to incorporate a protective effect of chloride and, if so, to what degree should it be reflected in the Ag BLM? Factors to consider in an effort to answer these questions are discussed below.

Evidence for the absence of a protective effect of Cl^- against Ag toxicity, for some species, is provided by a number of studies. For example, Bury and co-workers (Bury et al., 1999a), exposed >42-day old *P. promelas*, acclimated to low ionic strength water, to Ag over a range of NaCl levels (0.068-1.43 mM NaCl; Cl^- varied from 2.4 to 50.6 mg/L and Na^+ ranged from 1.9 to 36.6 mg/L). The tests resulted in a qualitatively consistent increase in *P. promelas* LC50 in association with the increase in NaCl. However, the response was not statistically significant and could reasonably be attributed to the increase in Na^+ alone, rather than the increase in Cl^-. In contrast to the *P. promelas* results, a statistically significant protective response was evident for the rainbow trout that were co-exposed to Ag in the same test vessels that held *P. promelas* (Bury et al., 1999a). It has also been shown that the ionoregulatory response of eels to Ag (*Anguilla anguilla*; a fish that is relatively insensitive to Ag with an LC50 of ~34 μg/L in soft water), in contrast to that of rainbow trout, does not receive a significant benefit by an increase in Cl^- over a range of 0.01 to 1.2 mM Cl^- (0.4 to 42.6 mg/L Cl^-, added as KCl; Grosell et al., 2000). More recently, tests in which the Cl^- was increased by 1 mM (by adding 1.0 mM KCl or 0.5 mM $MgCl_2$) exhibited a non-statistically significant increase in LC50 (~10% and 15%) in two of the three fish species tested (*D. rerio* and 4-day old *P. promelas*, respectively). A decrease in LC50 (also not statistically significant) was observed for the third species tested (*F. heteroclitus*), regardless of whether the Cl^- was adjusted by adding KCl (a 69% decrease) or $MgCl_2$ (an 18% decrease).

A possible explanation for the apparent differences in the responses of *O. mykiss* and *P. promelas* to chloride can be inferred from the results of the physiological tests that were performed by Bielmyer and co-workers (2008). While Ag had previously been shown to inhibit both Na^+ and Cl^- uptake by *O. mykiss* (Morgan et al., 2004), it only inhibited Na^+ uptake by *P. promelas* and not Cl^- uptake (Bielmyer et al., 2008). Thus, the mitigation of the inhibition of Cl^- uptake that provides an advantage to *O. mykiss* is not similarly advantageous to *P. promelas* because their Cl^- uptake is not inhibited by Ag^+ to begin with. While this rationale helps to explain the apparent species-specific differences in Cl^- protection, it is based on the premise that the responses to Ag of Na^+ and Cl^- uptake by *P. promelas* (only Na^+ uptake is inhibited) differs from that by *O. mykiss* (both Na^+ and Cl^- uptake are inhibited). The data to support this premise should be qualified, however, because the Ag concentration at which Na^+ uptake by *P. promelas* was inhibited (30 μg/L) was considerably higher than the Ag concentration at which Cl^- uptake by *P. promelas* was not inhibited (10 μg/L). That is, it is possible that neither Na^+ nor Cl^- uptake would have been inhibited at 10 μg/L and/or that both would have been inhibited at 30 μg/L. Even so, the collective results summarized above would seem to support a conclusion that Cl^- protection against toxicity due to Ag should only be invoked on a species-specific basis in the BLM.

An alternative view of the Cl^- results and of the need for incorporating organism-specific Cl^- protection in the BLM is obtained by considering the data from the perspective of the BLM. As currently structured, the version of the Ag BLM applied herein (Wu et al., 2008) considers both the Ag^+ ion and the neutral AgCl complex to be bioavailable and to accumulate at or within the target organ. Both forms of *accumulated* Ag are equally toxic and contribute to the LA50. However, in accordance with the results of Bury and Hogstrand (2002) the effective affinity for accumulation of Ag^+ at the site of action is higher than that of AgCl. The net effect is that the dissolved Ag concentration needed to reach the LA50 must increase as the Cl^- concentration increases and the distribution of dissolved Ag shifts from Ag^+ to AgCl (Figure 2-1). Thus, even though both of these forms of Ag are bioavailable, this representation leads to a limited protective effect of Cl^- at low to moderate Cl^- levels (less than about 50 mg/L ~ 1.4 mM, but dependent on other water quality characteristics), a response that would be difficult to detect in most of the datasets considered to date. At higher Cl^- concentrations (>100 mg/L ~ 2.8 mM) the Ag accumulation at the biotic ligand is controlled by the accumulation of AgCl, with the contribution of Ag^+ becoming relatively low. For a fixed total dissolved Ag concentration the concentration of the AgCl complex approaches a maximum at sufficiently high Cl^- levels (i.e., AgCl ~100% of the dissolved Ag). At this point further increases in chloride would no longer provide an additional protective effect if it were not for the formation of $AgCl_2^-$, assumed to be toxicologically benign, which reduces the concentration of AgCl. The dissolved Ag must therefore increase still further if the decrease in AgCl is to be offset and the LA50 attained. Hence further increases in Cl^- are protective as a result of the formation of $AgCl_2$ (and to a lesser degree other higher order chloro-complexes).

The preceding representation is generally consistent with the available information pertaining to Ag chemistry, toxicity and accumulation. With respect to chemistry, the inorganic speciation of Ag, including the formation of chloro-complexes, is based on binding constants obtained from the NIST database, a generally accepted source for thermodynamic constants. The Ag-biota interactions are based on a calibration of the Ag BLM to toxicity data, with the overall agreement between measured and predicted effect levels providing an indication that the net effect of the complex set of interactions that is represented is reasonable (Wu et al., 2008; Naddy et al., 2008). The assumption that AgCl accumulates and that $AgCl_2^-$ and other higher order chloro-complexes do not (i.e., they are not bioavailable or are low in bioavailability) is a fundamental premise that is explicitly represented in the BLM. Electrostatic and molecular size considerations suggest that this is a reasonable assumption given that these relatively large and negatively charged chloro-complexes are expected to be less amenable than Ag^+ and AgCl to accumulate via an active transport process and/or by passive diffusion. This BLM representation leads to a reasonable fit of the ~3-hour gill Ag accumulation data of Janes and Playle (1995). It is also qualitatively consistent with 4-day accumulation results for grass shrimp, where there was a clear relationship of Ag accumulation with the concentration of the neutral AgCl form rather than with Ag^+ (which was very low in concentration) or $AgCl_2^-$ (Engel et al., 1981). Similarly, it is in agreement with the more recent results of Bury and Hogstrand (2002) who measured 3-hour whole body Ag accumulation rates by *O. mykiss* and Atlantic salmon (*Salmo salar*) yolk-sac fry over a range of Cl^- levels of 0.05-10 mM (Cl^- was varied by adding KCl and K^+ was held constant by adding KNO_3). In this latter case it was reported that the results were similar for both species of fish, with the bioavailability of the important forms of Ag that were present varying as follows: $Ag^+ > AgCl(aq) >> AgCl_2^-$. They also concluded that Ag^+ controlled Ag accumulation at Cl < 1 mM and that the decrease in Ag accumulation at Cl > 1 mM was due to the formation of $AgCl_2$.

In view of the preceding considerations that tend to support alternative views of the protectiveness of Cl^-, it is useful to compare the magnitude of the BLM-predicted Cl^- responses to the observed responses in the experimental datasets where Cl^- was systematically varied by incremental additions of various chloride salts. The comparisons are summarized in Table 5-1. Results for fish shown in the upper portion of the table and results for the relatively sensitive invertebrates are presented in the middle section of the table. The range of Cl^- levels tested (with the exception of *C. dubia*, four Cl^- levels were tested) is shown in the second column and the measured LC50 in the lowest Cl^- treatment (the LC50 of the reference water) in the third column. The ratio of the LC50 at the highest Cl^- level to the LC50 of the reference (i.e., low Cl^-) water, as calculated with either the model results or the data, is also shown. This ratio provides an indication of the magnitude of the Cl^- response over the range of test conditions.

Table 5-1. Comparison of BLM-predicted and Measured Ag LC50 Responses in Tests with Systematically Varied Chloride.

Test Organism	Cl^- Test Range (mM)	Control LC50 (µg/L)	High/Low LC50 Ratio Data	High/Low LC50 Ratio Model***	Data Source (1)	Notes:
Calibration Datasets for Fish						
O. mykiss	0.068 – 1.43*	7.43	3.44	2.56	1	
P. promelas (>42 day)	0.068 – 1.43*	6.67	1.30	3.35	1	
P. promelas (28 day)	0.126 – 1.94	19.8	1.22 or 1.40	2.12 Or 1.24	2	2nd value for alternate reference Cl LC50
P. promelas (4 day)	0.085 – 1.70	2.49	2.22	1.74	2	Average LC50 for 3 hardness levels (BLM w/o low hardness to isolate Cl effect from IS effect)
Calibration Datasets for Invertebrates						
D. magna	0.085 - 1.48 0.271 – 1.48	0.93** 0.79**	1.34 1.58	1.51 1.37	2 2	Average for 2 hardness levels with alternative Cl reference LC50s
D. pulex	0.040 – 1.50	0.48	2.77	1.90	3	Average of 2 reference test results
C. dubia	0.124 – 8.20*	0.25	6.47	3.91	4	
Supplemental Fish and Invertebrate Data						
P. promelas (28 day)	0.123 & 0.323*	7.8 (6.8 – 8.9)	0.64	~1	5	
P. promelas (4 day)	0.01 – 1.0	2.32 2.32 – 2.32	1.15	1.56	6	
D. rerio	0.01 – 1.0	10.3 ~6.0 – 14.0	1.10	--	6	
F. heteroclitus	0.01 – 1.0 0.01 – 0.5**	3.88 (0.0 – 7.0)	0.31 0.82	-- --	6 6	
D. magna	0.050 – 1.00	0.47	2.55	1.62	3	

* NaCl was added rather than KCl. (model results reflect a Cl^- response plus a Na^+ response)

** $MgCl_2$ was added rather than KCl.

*** BLM-calculated LC50 based with 2 nmol / mg DOC.

1 Data Sources: (1) Bury et al., 1999a; (2) Karen et al., 1999; (3) Bury and Hogstrand, 2002; (4) Naddy et al., 2008; (5) Erickson et al., 1998; (6) Bielmyer et al., 2008

Note that some of the test results of Table 5-1 are based on nominal concentrations. To the degree this is the case, the comparisons which follow should be considered preliminary. With this in mind, inspection of the results shows that, in almost all cases, the Cl^- response predicted by the BLM is less than the Cl^- response that was observed. The most notable exceptions are the results for 42-day and 28-day *P. promelas*, the larger size and older age classes that are relatively insensitive to Ag (cf. the reference water LC50 and Figure 3-4). In contrast to these results the BLM-predicted Cl^- response for the more sensitive age class, 4-day old *P. promelas*, is somewhat less than the response indicated by the data (ratios of 1.74 versus 2.22, respectively).

The reason for the apparent age-related Cl^- response of *P. promelas* is not clear, but could plausibly be due to age-related differences in the mechanisms of ionoregulatory control that are operative (e.g., reliance on cutaneous versus branchial regulation) or to the increasing duration of the acclimation period with increasing age. The BLM also tends to predict less of a Cl^- response than is indicated by the invertebrate data. There is one marginal exception to this finding, the *D. magna* results of Karen et al. (1999). The BLM simulation of these test results leads to a greater increase in LC50 than is indicated by the data (Cl^- response ratios of 1.51 versus 1.34 over a Cl^- range of 0.085-1.48 mM) if the low Cl^- treatment is used as the LC50 reference level (0.93 µg/L). However, the measured LC50 of 0.79 µg/L was lower at the next higher Cl^- level tested (0.271 mM) and if this LC50 is used to evaluate the Cl^- response, the BLM response is lower than, but still in good agreement with, the response indicated by the data (ratios of 1.37 and 1.58, respectively).

In addition to the preceding results several other datasets that included toxicity tests in a reference lab water and for a single Cl^- treatment are available for consideration. With regard to fish, Erickson et al. (1998) reported that increasing the Cl^- from 0.123 to 0.323 mM by adding NaCl to the Lake Superior test water resulted in a *decrease* in the 28-day *P. promelas* LC50 from 7.8 to 5.0 µg/L. While the decrease was reported to be statistically significant the small increase in Cl^- level (0.2 mM, from 4.4 to 11.5 mg/L Cl^-) would limit the degree of any Cl^- response that might be operative and would make it difficult to differentiate it from normal test variability. As currently structured the BLM would not predict a decrease in LC50 as a result of the addition of NaCl to low DOC water and, for these particular test conditions, a slight increase in LC50 was predicted. These results should be considered with the results of Bielmyer and co-workers (Bielmyer et al., 2008), discussed above, where the measured *P. promelas* LC50 increased by ~15% (not statistically significant) when Cl^- was increased from 0.01 to 1.0 mM (added as KCl). The BLM predicted a greater than observed increase in LC50 for this particular set of experiments (56%), but one that is consistent with the underlying variability of the data. Bielmyer and colleagues also observed a 10% increase in the LC50 for *D. rerio* (not significant; KCl added) and a decrease in LC50 in two tests with *F. heteroclitus* (not significant; KCl or $MgCl_2$ added). BLM results are not available for these latter two fish species. Finally, Bury and Hogstrand (2002) also considered the effect of Cl^- on the toxicity of Ag to *D. magna*. Increasing the Cl^- from 0.05 to 1.0 mM by adding KCl resulted in a significant 2.55-fold increase in the LC50, from 0.47 to 1.20 µg/L. While the BLM response to chloride is somewhat less than the data, it also predicts an increase in LC50 of 1.62-fold for this set of test conditions.

One additional factor to consider with regard to the protectiveness of Cl^- against Ag toxicity warrants discussion. Specifically, the Cl^- level was in most cases varied by the addition of as much as 1.0 to 1.9 mM KCl (see Table 5-1), a method that necessarily confounds the interpretation of the results. The reason is that metals such as Cu and Ag are toxic to aquatic organisms via disruption of ionoregulation and, more specifically, inhibition of Na^+ and/or Cl^- uptake (Bianchini et al., 2002b; Bianchini and Wood, 2003; Grosell et al., 2002; Morgan et al., 2004). This leads to a progressive loss of these ions and, if sufficiently severe, has lethal consequences. For example, ~30% loss of plasma Na^+ has been shown to be lethal to fish (McDonald et al., 1980; Webb and Wood, 1998; Grosell et al., 2002) while somewhat higher losses of whole body Na^+ are lethal to *Daphnia* (Holm-Jensen, 1948). The transport of Na^+ across the branchial epithelium and into the organism is dependent upon the NKA enzyme system (Wood et al., 1996; Morgan et al., 1997) and, as originally shown in the seminal investigations of Jens Skou, NKA activity is closely tied to the extracellular fluid (ECF, which includes the plasma) and intracellular fluid (ICF, the cytosol) concentrations of both Na^+ and K^+ (Skou, 1957). Further, elevated levels of K^+ in water as low as 0.5 mM have been shown to perturb ECF levels (and presumably ICF levels) of both Na^+ and K^+, at least in the freshwater zebra mussel *Dreissena polymorpha* (a 3-fold increase in plasma K^+ within 1 day and 11% and 24% decreases in plasma Na^+ within 2 and 7 days, respectively; Wilcox and Dietz, 1995). It would seem reasonable to expect, then, that elevating the ambient K^+ level to 1.0 to 1.9 mM, as was done in many of the Cl^- tests with Ag, could potentially exacerbate the expected adverse ionoregulatory effects caused by exposure to Ag alone, partially if not entirely offsetting any potential benefit of the added Cl^- that might otherwise have been observed.

The lethal effect levels of K^+ alone have in some instances been found to be relatively high (typically >10 mM K^+; Mount et al., 1997). However, evidence that much lower K^+ levels will exacerbate the toxicity of an ionoregulatory inhibitor such as Ag is provided by test results with *P. promelas* exposed to Cu (Erickson et al., 1996), a metal having a mechanism of toxicity that is similar to Ag (Lauren and McDonald, 1987; Morgan et al., 1997). That is, Erickson and co-workers showed that the Cu LC50 was reduced by a factor of 2.2 to 3.6 relative to the control when the level of KCl was increased to 2.0 mM or higher (same Na^+ in control and KCl treatments; lower levels of KCl were not tested) (Erickson et al., 1996). This was in contrast to the 1.5 to 3.0-fold increase in Cu LC50 that was observed when 1 to 4 mM NaCl was added to the same test water. Thus, in view of the foregoing results, the potential for elevated K^+ levels to have reduced the protective effect of Cl^- against Ag toxicity in many of the datasets summarized in Table 5-1 should be viewed as a realistic possibility. Conversely, in cases where Cl^- was increased by adding NaCl there may have been a benefit associated with the increase in Na^+, such that the interpretation of the effect of Cl is somewhat clouded in these datasets as well. While the BLM provides a tool for attempting to quantify the magnitudes of the independent effects, future investigators should consider use of some other accompanying cation (e.g., addition of $MgCl_2$ or $CaCl_2$ to adjust Cl^-), with appropriate controls, to minimize the aforementioned complications.

5.2.5 Overall Consistency and Applicability of the BLM

As described in the preceding chapters the model calibration effort has involved a number of exploratory analyses that were completed in the process of evaluating different hypotheses that were proposed to explain the observations. Some of the areas that were investigated were discussed above and included the effects of ionic strength, age and size, chloride, and chemical kinetics on silver bioavailability and toxicity. Even though numerous sensitivity analyses have been considered, some of which involved adjustments to important model parameter values, there are actually only a few such variations that were actually used to produce the model results summarized herein. These model parameter values are summarized in Table 3-1 (chemical parameter values) and Table 3-2 (biotic ligand parameter values, including LA50s). Most of the model parameter values were consistent across all model simulations. An exception is that the LA50 was organism specific, an approach that has been used in the BLM since its original development and presentation to the U.S. EPA's SAB. Two notable variations that have been used in the case of *P. promelas* are that the LA50 was adjusted to account for differences in age- or size-related sensitivity to silver (analogous to adjusting the LA50 for differences in species sensitivity) and also to account for the apparent effect of low ionic strength waters on the sensitivity to silver of older age fish (the BLM simulation results are presented on Figure 4-7). The related LA50 relationships are included in Table 3-2. However, for implementation purposes, it is proposed that only the most sensitive age class, the 4-day old fish, be used for assessment purposes. Use of this sensitive age class not only provides a conservative basis for an effects assessment of WQC derivation purposes, but it also eliminates the need for varying the LA50 to address either age effects or ionic strength effects.

There are two additional model variants that were investigated over the course of model development that are discussed in the report. These were related to the initial natural water model testing results (summarized in Bielmyer et al., 2007). These BLM simulation results did not include the specification of 2 nmol/mg DOC of Ag^+ binding sites (log K = 12.0) that is represented in the current version of the BLM. Those early results also included competition between Ag^+ and other cationic metals for complexation by CRS. It was subsequently concluded on the basis of thermodynamic considerations and other related chemistry investigations (see Bell and Kramer, 1999, for a review) that Ag^+ would successfully displace these other metals such that competitive interactions could be neglected. With the exception of these interim results (displayed on Figure 4-2), all other BLM results presented herein were obtained with a single set of model parameter values used to represent Ag chemistry (Table 3-1), including 2 nmol/mg DOC Ag binding sites and without competition between Ag^+ and other metals.

The following approach is proposed for the near-term for implementation of the Ag BLM:

- Age Effects- Data for the most sensitive age class (e.g., 4-day old FHM) should be used for *P. promelas*. This will obviate the need for an empirical adjustment for age or size.
- Ionic Strength Effects- Use the relatively sensitive 4-day old *P. promelas* data or model results for assessment purposes (effects assessment, WQC derivation, etc.) will also eliminate the need to account for ionic strength on *P. promelas* sensitivity. If warranted, future refinements could address this matter in further detail, as new information and an improved understanding of the situation evolves.
- Chloride Protection- The Ag BLM currently provides a reasonable representation of the presently available data. Again, refinements could be implemented as called for by new information.
- Kinetic Effects- The BLM should continue to be applied as a chemical equilibrium-based model. Situations where kinetic effects are expected to be important should be addressed on a case-by-case basis. A more refined modeling approach that considers complexation kinetics could also be developed.

Note that, with the possible exception of the level of chloride protection that is represented in the BLM, the preceding topics are not necessarily silver- and/or BLM-related. As such, they would also need to be addressed in the context of a hardness-based WQC as well.

5.2.6 Are Increasing Concentrations of Competing Cations and Complexing Ligands Protective Against Ag Toxicity?

A question that is commonly asked, in a variety of forms, is the following: "Are increasing concentrations of cations (e.g., H^+, Na^+ and Ca^{2+}) and complexing ligands (e.g., Cl^-) protective of Ag toxicity?" Use of the BLM, a model that explicitly considers these interactions, provides a unique opportunity to explore the answer to this question. As it turns out, the question is not necessarily as straight forward to answer as one might expect. Three simple examples will illustrate why this is the case. First, consider the effect of pH. It appears from the experimental data that pH will have a limited effect on Ag bioavailability and toxicity in low DOC water, because pH has an inconsequential effect on Ag speciation over a typical environmental pH range. However, when DOC is present at an elevated level, an increase in pH increases the effective complexation capacity of the NOM for Ag^+ (due to deprotonation of NOM binding sites), thereby leading to a decrease in toxicity with increasing pH. As a second example, one that is contrary to the response of the standard hardness equation, the BLM also indicates that an increase in Ca^{2+} can lead to an increase in Ag availability and toxicity if the DOC is elevated, because the Ca^{2+} ions may bind to NOM sites that would otherwise be available to bind Ag^+. Finally, as a third example, while Cl^- might normally be expected to provide a limited level of protection, an increase in Cl^- concentration can, at least in principle, lead to a decrease in LC50. This would be expected to occur if the level of organic matter (OM) is sufficiently high such that an increase in Cl^- increases the AgCl fraction, which is bioavailable, at the same time that the Ag-NOM fraction, a complex that is considered to be low in bioavailability, decreases.

The preceding examples illustrate how the BLM can be used to highlight (and quantify) some less than obvious but conceivable responses to Ag. In doing so, it provides insight into the net effect of what would otherwise be a set of difficult to decipher abiotic and biotic interactions. While the model can explicitly represent such interactions it must also subsume any physiological interactions that occur into the overall calibration of the BLM to toxicity data. As a result of this, the secondary import of many of the chemical interactions that occur, and the inherent variability of toxicity test results, it is somewhat difficult to detect such responses with standard testing protocols. The BLM represents a tool that can be used to identify such nuances in the data and to tailor the design of future laboratory investigations that are carried out to explore their validity. The process is an iterative one, given that the data can be used to further refine the model, and the model in turn to refine the experimental design.

In summary then, the answer to the original question: "Are increasing concentrations of competing cations and complexing ligands protective of Ag toxicity?", is not nearly as clearly defined as one would hope. The answer is "It depends… ".

5.3 Progress Towards the Achievement of Management Objectives

As stated in Chapter 1.0, the WERF Ag BLM investigations represented one important aspect of a considerably broader effort to understand the fate and effects of Ag in the environment. These Ag-related investigations were coordinated such that the collective information would help to satisfy a number of important management objectives, objectives that emanated from both the regulatory and regulated communities. These management objectives included the eventual derivation of scientifically defensible acute and chronic WQC for Ag in both freshwater and saltwater environments. It was important that these WQC be able to account for the effects of site-specific conditions in natural waters. Further, in order to meet the needs of both the regulatory (e.g., permit writers) and regulated (i.e., dischargers) end-users it was important that the models to be used to achieve these objectives not be overly complex.

Considerable progress has been made toward the achievement of the aforementioned management objectives via the WERF Ag BLM effort in combination with a number of related industry-sponsored research initiatives (including but not limited to Ag, Cu, Ni and Zn initiatives), the WERF Ag and Cu BLM projects, and U.S. EPA efforts. These efforts have led to the development of a user-friendly version of the BLM that can be used to evaluate BLM-based site-specific WQC. The recently released BLM-based WQC document for Cu represents the most tangible result to date in this regard (U.S. EPA, 2007). With the completion of the WERF Ag BLM studies, in combination with the parallel investigations of the broader Ag research and development program (Paquin et al., 2007), EPA recently initiated efforts to incorporate the Ag

BLM into an updated ambient WQC for Ag. While further research is needed, particularly with regard to advancing the state of the art in the derivation of technically defensible marine and chronic WQC, an extensive body of information now exists that will be of use in achieving these objectives.

5.4 Conclusions and Recommendations

The following conclusions are arrived at on the basis of the results of the WERF Ag BLM project:

- Based on the consistency of the BLM speciation results with the available Ag speciation data, and the ability of the model to reproduce the organism response to Ag that is observed in the majority of the toxicity tests, the Ag BLM seems to provide a reasonable estimate of Ag speciation in natural waters.
- Consistent with U.S. EPA SAB recommendations, the toxicity database that is available for Ag BLM development purposes has been expanded considerably. It now includes an extensive set of data for *C. dubia*, as well as additional data for *P. promelas*, with both sets of test results reflecting an expanded range of water quality characteristics.
- The updated acute Ag BLM should serve as an appropriate tool for incorporation into an updated WQC document for Ag. This BLM-based WQC should be amenable to updating as the current understanding of the bioavailability and effects of Ag (and other metals) on aquatic organisms continues to advance.

Initial efforts to validate the Ag BLM with fish and invertebrate toxicity data that were obtained with natural water samples met with mixed results. The *P. promelas* analyses in particular were initially unsuccessful and highlighted the need to refine the BLM to account for an apparent effect of low ionic strength water on the sensitivity to Ag of older age *P. promelas*.

- The updated model is suitable for use in predicting LC50 values for both *P. promelas* and *C. dubia* over a wide range of water quality characteristics. The results for the relatively sensitive *C. dubia* were particularly informative in that they included data that demonstrated the ability of the model to predict the protection against toxicity due to Ag that is provided by elevated levels of CRS in natural waters. While ionic strength effects complicate the interpretation of results for fathead minnows, it appears that data for the most sensitive age class tested to date (4-day old *P. promelas*) is not affected by this factor.

Given that the available data are subject to alternative interpretations, the general ability of Cl^- to protect against Ag toxicity across species and how to best represent Cl^- protection in the BLM would benefit from further clarification. One view is that the absence of a definitive (i.e., statistically significant) response to chloride in some of the experimental datasets warrants that Cl^- protection be invoked on a species-specific basis. Alternatively, in view of the general consistency between BLM results and the current body of information, use of a generic model of Cl^- protection is reasonable. Additional data will help to resolve this uncertainty. The decision to be made at present is whether to proceed with the BLM in its current form, or to wait until future data become available for consideration. A relatively conservative approach would be to continue to use the Ag BLM, but without an allowance for Cl^- protection. A disadvantage of this approach is that it would necessitate a recalibration of the BLM to the available data and doing so could lead to the unintended consequence that compensatory artifacts are introduced into the BLM calibration. Alternatively, given the tendency of the currently calibrated version of the BLM to predict a protective response to Cl^- that is less than is indicated by the majority of the datasets analyzed (including the 4-day old *P. promelas* age class), it could continue to be used in its present form with additional refinements being incorporated, as appropriate, as new data become available for consideration.

As described herein, the Ag investigations that were carried out have helped to advance that state of understanding of the bioavailability and toxicity of Ag in aquatic environments. As is normally the situation with this type of research effort, however, there are a few key technical areas that would benefit from further study: More specifically it is recommended that:

- In light of the failure of some studies to have demonstrated a statistically significant protective effect of chloride against Ag toxicity, the uncertainty associated with standard test protocols that have been routinely used in the past, and the current representation of a limited protective benefit in the BLM, toxicity tests incorporating a more robust experimental design should be carried out. The standard test design could be enhanced by including: an increased level of replication, a chemistry component that considers Ag speciation, side-by-side comparisons where Cl^- is added with different cations (e.g., Na^+, K^+, Mg^{2+}, Ca^{2+}), physiological measurements of ionoregulatory end-points (e.g., plasma ion levels and one way ion fluxes), confirmatory testing by an independent laboratory, and modeling.
- The effect of acclimation conditions and of low IS water on organism sensitivity to metals are areas that require further exploration. Given the generic need for this type of information, these studies should include tests with both fish and invertebrate species and should be performed not only with Ag, but with other metals as well. Multi-generational studies, while also informative to complete, would not be necessary for further development of an acute BLM. The study design should include the following components: chemistry (i.e., a careful characterization of metal speciation), physiology, toxicology and modeling.

APPENDIX A

LIST OF PROJECT MANUSCRIPTS

Bielmyer, G.K., M. Grosell, P.R. Paquin, R. Mathew, K.B. Wu, K.V. Brix, and R. Santore, 2007. "Validation Study of the Acute Biotic Ligand Model for Silvér," *Environmental Toxicology and Chemistry*, 26(10): 2241-2246.

Bielmyer, G.K., K.V. Brix, and M. Grosell, 2008. "Is Chloride Protection against Silver Toxicity Due to Water Chemistry," *Aquatic Toxicology*, In Press.

Naddy, R.B., G.R. McNerney, W.A. Stubblefield, R.A. Bell, K.B. Wu, P.R. Paquin, and R. Santore, 2008. "Influence of Varying Water Quality Parameters on the Acute Toxicity of Silver to the Freshwater Cladoceran, *Ceriodaphnia dubia*," in preparation.

Santore, R., R. Bell, R. Mathew, P.R. Paquin, D.S. Smith, K. Wu, and D.M. Di Toro, 2008. "Development of an Acute BLM for Ag: Ag Speciation in Natural Waters with Organic Matter Interactions," in preparation.

Smith, D.S., R.A. Bell, and J.R. Kramer, 2002. "Metal Speciation in Natural Waters with Emphasis on Reduced Sulfur Groups as Strong Metal Binding Sites," Special Issue: The Biotic Ligand Model for Metals–Current Research, Future Directions, Regulatory Implications, *Comparative Biochemistry and Physiology*, Part C 133(1-2): 65-74.

Smith, D.S., R.A. Bell, J. Valliant, and J.R. Kramer, 2004. "Determination of Strong Ligand Sites in Sewage Effluent-Impacted Waters by Competitive Ligand Titration with Silver," *Environmental Science and Technology*, 38(7): 2120-2125.

Wu, K.-B., R. Mathew, G. Bielmyer, K. Brix, R. Naddy, R. Bell, M.Grosell, P.R. Paquin, R. Santore, and D.M. Di Toro, 2008. "Development of an Acute Biotic Ligand Model for Silver: Application to Fish and Invertebrates," in preparation.

APPENDIX B

LABORATORY AND NATURAL WATER TEST RESULTS: DATA SUMMARY

Table B-1. Laboratory Water *Ceriodaphnia dubia* Test Results (Naddy et al., 2008).

Source	Test Date	Hardness (mg/L as $CaCO_3$)	Alkalinity (mg/L as $CaCO_3$)	pH	Cl (µM)	Ca (µM)	Mg (µM)	K (µM)	Na (µM)	SO4 (µM)	DOC (mg/L)	DOC (µM)	CRS nM S2-	Lt nM Ag	Total LC50 (µg/L)	Dissolved LC50 (µg/L)
Hardness/Alkalinity studies																
100 / 50		97	47	7.6	79	624	341	41	1	937	1	83			0.38	0.38
100 / 100		99	101	7.7	138	624	366	65	2	999	0.6	50			0.16	0.15
160 / 100		148	99	7.6	158	948	535	70	2	1562	0.7	58			0.17	0.14
300 / 100		278	96	7.7	197	1796	987	126	2	2915	0.8	67			0.17	0.16
200 / 200		186	200	7.9	121	1198	658	100	3	1874	1.2	100			0.21	0.21
320 / 200		295	198	7.8	130	1921	1029	146	3	3123	0.7	58			0.21	0.18
600 / 200		547	200	7.9	313	3493	1975	262	3	5934	0.6	50			0.14	0.11
NaCl Studies																
Low		88	92	7.8	124	349	535	43	2	958	0.4	33			0.31	0.25
Medium		88	94	8.0	564	349	535	43	2	947	0.9	75			0.37	0.33
High		88	98	8.0	8208	349	535	57	7	947	0.5	42			1.63	1.63
Na_2SO_4 Studies																
Low		88	98	8.0	85	349	535	44	2	937	0.9	75			0.56	0.52
Medium		88	99	8.0	90	349	535	47	2	1145	0.6	50			0.53	0.50
High		84	98	8.0	71	349	494	61	7	4997	0.5	42			0.74	0.69
Low Na Studies																
Low		50	8	7.0	65	324	173	23	0	500	0.5	42			0.20	0.20
Medium		49	24	7.7	62	324	165	23	0	500	0.5	42			0.32	0.30
High		47	60	8.2	54	299	173	28	1	500	0.5	42			0.35	0.33
pH studies																
100 – ambient		97	95	8.3	141	624	341	43	2	916	2	167			0.45	0.42
100 – CO2		97	96	7.4	141	624	341	43	2	916	2	167			0.26	0.25
200 – ambient		89	190	8.3	59	574	321	43	3	999	1.1	92			0.22	0.18
200 – CO2		89	186	7.1	59	574	321	43	3	999	1.1	92			0.52	0.45
NOM studies																
Recon match		197	49	8.1	4005	1272	699	118	4	1874	1.6	133			0.61	0.58
Low		207	54	8.1	6967	1372	699	180	6	1353	3.5	292	33		> 2.88	> 2.76
Medium		192	54	8.1	7164	1347	576	229	6	625	6.9	575	65		4.90	4.68
High		328	84	8.2	6798	2295	987	360	5	1041	10.5	875	109		5.61	5.39
Site water hardness match																
Desjardins	9/18-20/01	188	63	7.7	3921	1223	658	103	4	1978	0.7	58				0.36
Desjardins	1/8-10/02	212	99	8.2	6459	1422	699	129	7	2290	0.1	8				0.59
Desjardins	4/17-19/02	300	171	8.4	5387	1846	1152	154	9	3123	1.9	158				1.02
Grand River	5/22-24/02	243	207	8.5	1503	1522	905	119	6	2707	0.7	58				0.43
Rochester	7/31-8/02	170	208	8.1	2570	1123	576	108	6	1666	0.3	25				0.87
Black River	8/29-31/02	17	9	7.6	161	107	62	16	0	177	0.5	42				0.16

Table B-2. Dataset 1: Natural Water *Ceriodaphnia dubia* Test Results.

Source	Test Date	Hardness	Alkalinity	Initial pH	Cl	Ca	Mg	K	Na	SO4	DOC	DOC	CRS	Lt	Total LC50	Dissolved LC50
		(mg/L as $CaCO_3$)			(µM)	(µM)	(µM)	(µM)	(µM)	(µM)	(mg/L)	(µM)	nM S2-	nM Ag	(µg/L)	(µg/L)
Raw Water Samples																
Desjardins	9/18-20/01	200	61	7.1	3949	1248	576	358	3175	666	7.0	583	67		4.93	3.92
Desjardins	1/8-10/02	232	98	7.2	6347	1422	658	332	4785	760	5.9	492	141		11.1	9.50
Desjardins	4/17-19/02	318	172	7.2	5388	2246	987	215	4785	906	5.8	483	188		10.8	9.91
Grand River	5/22-24/02	272	217	8.5	1574	1871	864	61	1392	323	5.2	433	78	64	1.60	1.40
Rochester	7/31-8/2/02	174	221	7.6	2759	1347	576	307	2479	406	12.0	1000	211	200	6.79	5.24
Black River	8/29-31/02	14	13	7.2	65	90	53	<26	69.6	60	3.0	250	50		1.22	1.02
	Minimum	14	13	7.1	65	90	53	<26	69.6	60	3.0	250	50	64	1.22	1.02
	Maximum	318	221	8.5	6347	2246	987	358	4785	906	12	1000	211	200	11.1	9.91
Filtered Water Samples																
Desjardins	1/8-10/02	238	96	7.2	6234	1472	658	332	4785	750	5.1	425	124		11.0	12.1
Desjardins	4/17-19/02	314	175	7.5	5501	2246	987	215	4785	895	5.3	442	147		9.37	9.25
Grand River	5/22-24/02	270	218	8.3	1520	1871	864	59	1392	323	5.4	450	36		0.97	0.99
Rochester	8/7-9/02	178	226	7.7	2773	1347	576	307	2479	406	10.9	908	202		3.53	3.46
Black River	8/29-31/02	14	13	7.2	71	87	53	<26	65	60	3.2	267	48		1.06	0.90
	Minimum	14	13	7.2	71	87	53	<26	65	60	3.2	267	36		0.97	0.90
	Maximum	314	226	8.3	6234	2246	987	332	4785	895	10.9	908	202		11.0	12.1
RO Samples																
Desjardins	1/17-19/02	240	90	7.8	6601	1397	782	256	5220	1145	6.7	558	122		13.2	8.85
Desjardins	5/9-11/02	355	166	8.0	6544	2121	1604	194	6525	1978	7.7	642	142		2.52	2.53
Grand River	5/29-31/02	260	204	8.1	2477	1597	1316	79	3306	1031	6.6	550			1.26	1.16
Rochester	8/7-9/02	182	218	7.8	2827	1223	740	307	3045	697	10.5	875			1.63	1.65
Black River	9/3-5/02	16	12	7.1	76	100	62	<26	78	78	3.4	283			0.56	0.54
	Minimum	16	12	7.1	76	100	62	<26	78	78	3.4	283	122		0.56	0.54
	Maximum	355	218	8.1	6601	2121	1604	307	6525	1978	10.5	875	142		13.2	8.85
RE Samples																
Desjardins	1/17-19/02	234	94	7.6	7165	1497	864	384	6090	1249	6.8	567	98		8.61	7.95
Desjardins	5/9-11/02	385	160	8.1	8237	2221	1728	332	8699	2082	7.2	600	163		7.28	4.07
Grand River	5/29-31/02	260	210	8.1	1913	2270	1399	79	3610	1249	6.2	517			0.73	0.66
	Minimum	234	94	7.6	1913	1497	863.8	79.28	3610	1249	6.2	516.7	98		0.73	0.66
	Maximum	385	210	8.1	8237	2270	1728	383.6	8699	2082	7.2	600	163		8.61	7.95

RO Test water was reconstituted from NOM concentrate obtained by reverse osmosis

RE Test water was reconstituted from NOM concentrate obtained by rotary evaporation

Table B-3. Dataset 2: Natural Water Test Results (Bielmyer et al., 2007).

Source	Hardness (mg/L as CaCO3)	Total CO2 (µM)	Initial pH	Cl (µM)	Ca (µM)	Mg (µM)	K (µM)	Na (µM)	SO4 (µM)	DOC (mg/L)	DOC (µM)	CRS nM S2-	*P. promelas* Dissolved LC50 (µg/L) LC50	*C. dubia* Dissolved LC50 (µg/L) 95% CI	LC50	95% CI
UM Reference 1	87	888	7.74	1352	512	251	105	1076	121	4.4	370	1.0	3.37	3.02, 3.75		
UM Reference 2	63	730	8.07	1189	451	175	100	928	130	4.5	379	0.7			0.76	0.53, 1.09
Horsetooth Res: Unacclimated	34	320	7.75	138	166	117	7	184	50	3.9	321	1.5	5.23	4.40, 6.23	1.28	0.63, 2.60
Horsetooth Res: Acclimated													2.44	1.21, 4.89		
Green Cove Springs	132	1300	8.14	210	682	628	8	257	520	1.8	148	3.6	6.02	3.63, 10.0	0.34	0.19, 0.61
Desjardins	190	2050	8.10	4358	1485	761	294	4473	625	5.7	472	49.0	20.9	19.2, 22.7	3.24	1.98, 5.28
Everglades*	197	3940	8.04	796	1719	250	8	752	52	11.5	960	21.0	44	42.5, 45.5	1.45	1.19, 1.77
Miller Creek*	144	2950	7.84	105	600	693	11	520	24	4.3	362	21.0	16	12.6, 20.4	4.24	3.36, 5.34
Ringwood*	40	500	7.20	594	284	109	6	456	83	6.2	513	5.0	20.2	18.3, 21.9	1.15	0.58, 2.31
St. Louis Bay, MN 1	110	1400	7.79	187	508	338	32	208	78	12.4	1037	19.0	44.1	UC		
St. Louis Bay, MN 2	84	1340	7.82	204	499	298	44	226	94	12.0	1000	26.5			9.52	7.7, 11.77
Trout Lake, ON: Unacclimated	11	100	6.84	183	78.6	35	27	138	48	5.0	418	1.0	2.49	1.88, 3.30	1.69	1.30, 2.20
Trout Lake, ON: Acclimated													2.96	2.41, 3.64		
10% UM Ref: Unacclimated	13	40	6.17	173	39.8	23	6	123	14	2.4	202	6.0	1.99	1.70, 2.34		
10% UM Ref: Acclimated													2.18	1.69, 2.81		

* = Filtered Sample (0.45 µm)

REFERENCES

Adams, N.W.H. and J.R. Kramer, 1998. "Reactivity of Ag+ Ion with Thiol Ligands in the Presence of Iron Sulfide," *Environmental Toxicology and Chemistry*, 17(4): 625-629.

Adams, N.W.H. and J.R. Kramer, 1999. "Silver Speciation in Wastewater Effluent, Surface Waters and Pore Waters," *Environmental Toxicology and Chemistry*," 18(12): 2667-2673.

Andren, A.W. and T. Bober, eds., 2002. *Silver in the Environment: Transport, Fate and Effects*, Research Findings of the Argentum International Conference Series, 1993-2000, Pensacola, FL, USA: Society of Environmental Toxicology and Chemistry (SETAC), 169 pp.

ASTM (American Society for Testing and Materials), 2001. "Standard Guide for Conducting Acute Toxicity Tests with Fishes, Macroinvertebrates, and Amphibians," E729-96, in Annual Book of ASTM Standards, Vol 11.05. Philadelphia, PA, pp. 175-196.

Bell, R.A. and J.R. Kramer, 1999. "Structural Chemistry and Geochemistry of Silver-Sulfur Compounds: Critical Review," *Environmental Toxicology and Chemistry*, 18(1): 9-22.

Bianchini, A., K.C. Bowles, C.J. Brauner, J.W. Gorsuch, J.R. Kramer, and C.M. Wood, 2002a. "Evaluation of the Effect of Reactive Sulfide on the Acute Toxicity of Silver(I) to *Daphnia magna*, Part 2: Toxicity Results," *Environmental Toxicology and Chemistry,* 21(6): 1294–1300.

Bianchini, A., M. Grosell, S.M. Gregory, and C.M. Wood, 2002b. "Acute Silver Toxicity in Aquatic Animals is a Function of Sodium Uptake Rate," *Environmental Science and Technology*, 36(8): 1763-1766.

Bianchini, A. and C.M. Wood, 2003. "Mechanism of Acute Silver Toxicity in *Daphnia magna*," *Environmental Toxicology and Chemistry*, 22(6): 1361-1367.

Bielmyer, G.K., M. Grosell, P.R. Paquin, R. Mathew, K.B. Wu, K.V. Brix, and R. Santore, 2007, "Validation Study of the Acute Biotic Ligand Model for Silver," *Environmental Toxicology and Chemistry*, 26(10): 2241-2246.

Bielmyer, G.K., K.V. Brix, and M. Grosell, 2008. "Is Chloride Protection against Silver Toxicity Due to Water Chemistry," *Aquatic Toxicology*, in press.

Bowles, K.C., R.A. Bell, M.J. Ernste, J.R. Kramer, H. Manolopoulos, and N. Ogden, 2002a. "Synthesis and Characterization of Metal Sulfide Clusters for Toxicological Studies," *Environmental Toxicology and Chemistry*, 21(4): 693-699.

Bowles, K.C., A. Bianchini, C.J. Brauner, J.R. Kramer, and C.M. Wood, 2002b. "Evaluation of the Effect of Reactive Sulfide on the Acute Toxicity of Silver (I) to *Daphnia magna*. Part 1:

Description of the Chemical System," *Environmental Toxicology and Chemistry*, 21(6): 1286-1293.

Bowles, K.C., M.J. Ernste, and J.R. Kramer, 2003. "Trace Sulfide Determination in Oxic Freshwaters," *Analytica Chimica Acta*, 477: 113-124.

Brix, K.V., R. Santore, D.G. Henderson, P.R. Paquin, J.W. Gorsuch, and W.J. Adams, 1999. "Evaluation of Water Effect Ratios for Copper and Silver Using the Biotic Ligand Model," platform presentation at the 1999 Annual SETAC Conference, Philadelphia, PA.

Brooks, B.W., J.K. Stanley, J.C. Winter, P.K. Turner, K.B. Wu, and T.W. La Point, 2004. "Laboratory and Field Responses to Cadmium: An Experimental Study in Effluent Dominated Stream Mesocosms," *Environmental Toxicology and Chemistry*, 23(4): 1057-1064.

Bury, N.R. and C.M. Wood, 1999. "Mechanism of Branchial Apical Silver Uptake by Rainbow Trout is via the Proton-Coupled Na^+ Channel," *American Journal of Physiology 277 (Regulatory Integrative Comparative Physiology* 46): R1385-R1391.

Bury, N.R., F. Galvez, and C.M. Wood, 1999a. "Effects of Chloride, Calcium and Dissolved Organic Carbon on Silver Toxicity: Comparison between Rainbow Trout and Fathead Minnows," *Environmental Toxicology and Chemistry*, 18(1): 56-62.

Bury, N.R., M. Grosell, A.K. Grover, and C.M. Wood, 1999b. "ATP-Dependent Silver Transport Across the Basolateral Membrane of Rainbow Trout Gills," *Toxicology and Applied Pharmacology*, 159: 1-8.

Bury, N.R. and C. Hogstrand, 2002. "Influence of Chloride and Metals on Silver Bioavailability to Atlantic Salmon (*Salmo salar*) and Rainbow Trout (*Oncorhynchus mykiss*) Yolk-sac Fry," *Environmental Science & Technology*, 36: 2884-2888.

Campbell, P.G.C., 1995. "Interactions Between Trace Metals and Aquatic Organisms: A Critique of the Free-ion Activity Model," *in Metal Speciation and Bioavailability in Aquatic Systems*, A. Tessier and D.R. Turner, eds., IUPAC, John Wiley and Sons, New York, 45-102.

Campbell, P.G.C., P.R. Paquin, W.J. Adams, K.V. Brix, D.R. Juberg, R.C. Playle, C.J. Ruffing, and R.S. Wentsel, 2002. "Risk Assessment," Chapter 4 in *Silver in the Environment: Transport, Fate and Effects*, Research Findings of the Argentum International Conference Series, 1993-2000, A.W. Andren and T.W. Bober, eds., Pensacola, FL, USA: Society of Environmental Toxicology and Chemistry (SETAC), pp. 97-139.

Davies, S.N. and R.C.M. DeWiest, 1966. *Hydrogeology*, John Wiley, New York.

De Schamphelaere, K.A.C. and C.R. Janssen, 2002a. "A Biotic Ligand Model Predicting Acute Copper Toxicity for *Daphnia magna*: the Effects of Calcium, Magnesium, Sodium, Potassium and pH," *Environmental Science and Technology*, 36(1): 48-54.

De Schamphelaere, K.A.C. and C.R. Janssen, 2004. "Development and Field Validation of a Biotic Ligand Model Predicting Chronic Copper Toxicity to *Daphnia magna*," *Environmental Toxicology and Chemistry*, 23(6): 1365-1375.

Di Toro, D.M., H.E. Allen, H.L. Bergman, J.S. Meyer, R.C. Santore, and P.R. Paquin, August 2000. "The Biotic Ligand Model: A Computational Approach for Assessing the Ecological Effects of Metals in Aquatic Systems," published by the International Copper Association, Ltd., Environmental Program, as part of its series on "*Copper in the Environment and Health*".

Di Toro, D.M., H.E. Allen, H.L. Bergman, J.S. Meyer, P.R. Paquin, and R.C. Santore, 2001. "A Biotic Ligand Model of the Acute Toxicity of Metals. I. Technical Basis," *Environmental Toxicology and Chemistry,* 20(10): 2383-2396.

Engel, D.W., W.G. Sunda, and B.A. Fowler, 1981. "Factors Affecting Trace Metal Uptake and Toxicity to Estuarine Organisms. I. Environmental Parameters," in *Biological Monitoring of Marine Pollutants*, F.J. Vernberg, A. Calabrese, F.P. Thurberg and W.B. Vernberg, eds., Academic Press, NY, USA, pp. 127-144.

Erickson, R.J., C.F. Kleiner, J.T. Fiandt, and T.L. Highland, 1997. "Effect of Acclimation Period on the Relationship of Acute Copper Toxicity to Water Hardness for Fathead Minnows," *Environmental Toxicology and Chemistry*, 16(4): 813-815.

Glover, C., R.C. Playle, and C.M. Wood, 2005a. "Heterogeneity of Natural Organic Matter Amelioration of Silver Toxicity to *Daphnia magna*: Effect of Source and Equilibration Time," *Environmental Toxicology and Chemistry*, 24(11): 2934-2940.

Glover, C., S.K. Sharma, and C.M. Wood, 2005b. "Heterogeneity in Physicochemical Properties Explains Differences in Silver Toxicity Amelioration by Natural Organic Matter to Daphnia magna," *Environmental Toxicology and Chemistry*, 24(11): 2941-2947.

Grosell, M., C. Nielsen, and A. Bianchini, 2002. "Sodium Turnover Rate Determines Sensitivity to Acute Copper and Silver Exposure in Freshwater Animals," Special Issue: The Biotic Ligand Model for Metals–Current Research, Future Directions, Regulatory Implications, *Comparative Biochemistry and Physiology*, Part C 133(1-2): 287-303.

Herrin, R.T., A.W. Andren, M.M. Shafer, and D.E. Armstrong, 2001. "Determination of Silver Speciation in Natural Waters. 2. Binding Strength of Silver Ligands in Surface Freshwaters," *Environmental Science and Technology*, 35:1959-1966.

Hollis, L., K. Burnison, and R.C. Playle, 1996. "Does the Age of Metal-Dissolved Organic Carbon Complexes Influence Binding of Metals to Fish Gills?" *Aquatic Toxicology*, 35: 253-264.

Holm-Jensen, I.B., 1948. "Osmotic Regulation in Daphnia magna under Physiological Conditions and in the Presence of Heavy Metals," *K. Dan Vidensk Selsk Biol. Medd.*, 20: 1-64.

HydroQual, 2001. "Review of Phase I Efforts to Develop a Biotic Ligand Model for Nickel," prepared for the Water Environment Research Foundation, Alexandria, VA.

HydroQual, 2002. "Review of Phase I Efforts to Develop a Biotic Ligand Model for Lead," draft report submitted to the International Lead and Zinc Organization.

HydroQual, 2003. "Phase I Development of a Biotic Ligand Model for Cadmium," draft report submitted to the Battelle Laboratories and the United States Environmental Protection Agency, Washington, DC.

Janes, N. and R.C. Playle, 1995. "Modeling Silver Binding to Gills of Rainbow Trout (*Oncorhynchus mykiss*)," *Environmental Toxicology and Chemistry*, 14(11): 1847-1858.

Karen, D.J., D.R. Ownby, B.L. Forsythe, T.P. Bills, T.W. La Point, G.B. Cobb, and S.J. Klaine, January 1999. "Influence of Water Quality on Silver Toxicity to Rainbow Trout, (*Oncorhynchus mykiss*), Fathead Minnows (*Pimephales promelas*), and Water Fleas (*Daphnia magna*)," *Environmental Toxicology and Chemistry*, 18(1): 63-70.

Kinniburgh, D.G., C.J. Milne, M.F. Benedetti, J.P. Pinheiro, J. Filius, L.K. Koopal, and W.H. Van Riemsdijk, 1996. "Metal Ion Binding by Humic Acid: Application of the NICA-Donnan Model," *Environmental Science and Technology*, 30(5): 1687-1698.

Kolts, J.M., C.J. Boese, and J.S. Meyer, 2006. "Acute Toxicity of Copper and Silver to *Ceriodaphnia dubia* in the Presence of Food," *Environmental Toxicology and Chemistry*, 25(7): 1831-1835.

Kramer, J.R., N.W.H. Adams, H. Manolopoulos, and P.V. Collins, 1999. "Silver at an Old Mining Camp, Cobalt, Ontario, Canada," *Environmental Toxicology and Chemistry*, 18(1): 23-29.

Kramer, J.R., R.A. Bell, and D.S. Smith, 2007. "Determination of Sulfide Ligands and Association with Natural Organic Matter," *Applied Geochemistry,* 22: 1606-1611.

Kuwabara, J.S. and G.W. Luther, III, 1993. "Dissolved Sulfides in the Oxic Water Column of San Francisco Bay, California," *Estuaries*, 16(3A): 567-573.

Lauren, D.J. and D.G. McDonald, 1987. "Acclimation to Copper by Rainbow Trout, *Salmo gairdneri*: Biochemistry," *Canadian Journal Fish. Aquat. Sci.* 44: 105-111.

Lemke, A.E., February 1981. "Interlaboratory Comparison Acute Testing Set," prepared for Office of Pesticides and Toxic Substances, Information Center, National Fisheries Contaminant Research Center, EPA 600/3-81-005.

Luther, G.W. III and D.T. Rickard, 2005. "Metal Sulfide Cluster Complexes and their Biogeochemical Importance in the Environment," *Journal of Nanoparticle Research*, 7:389-407.

Ma, H., S.D. Kim, D.K. Cha, and H.E. Allen, 1999. "Effect of Kinetics of Complexation by Humic Acid on the Toxicity of Copper to *Ceriodaphnia dubia*," *Environmental Toxicology and Chemistry*, 18(5): 828-837.

MacRae, R.K., D.E. Smith, N. Swoboda-Colberg, J.S. Meyer, and H.L. Bergman, 1999. "Copper Binding Affinity of Rainbow Trout (*Oncorhynchus mykiss)* and Brook Trout (*Salvelinus fontinalis)* Gills: Implications for Assessing Bioavailable Metal," *Environmental Toxicology and Chemistry*, 18(6): 1180-1189.

Mann, R.M., M.J. Ernste, R.A. Bell, J.R. Kramer, and C.M. Wood, 2004. "Evaluation of the Protective Effect of Reactive Sulfide on the Acute Toxicity of Silver to Rainbow Trout (*Oncorhynchus mykiss*)," *Environmental Toxicology and Chemistry*, 23(5): 1204-1210.

Martell, A.E., R.M. Smith, and R.J. Motekaitis, 2003. NIST Standard Reference Database 46 Version 7.0. NIST Critically Selected Stability Constants of Metal Complexes. NIST Standard Reference Data, Gathersburg, MD.

McDonald, D.G., H. Hobe, and C.M. Wood, 1980. "The Influence of Calcium on the Physiological Responses on the Rainbow Trout, *Salmo, gairdneri* to Low Environmental pH," *Journal of Experimental Biology*, 88: 109-131.

McGeer, J.C. and C.M. Wood, 1998. "Protective Effects of Water Cl^- on Physiological Responses to Waterborne Silver in Rainbow Trout," *Canadian Journal of Fisheries and Aquatic Sciences*, 55: 2447-2454.

McGeer, J.C., R.C. Playle, C.M. Wood, and F. Galvez., 2000. "A Physiologically Based Biotic Ligand Model for Predicting the Acute Toxicity of Waterborne Silver to Rainbow Trout in Freshwaters," *Environmental Science and Technology*, 34(19): 4199-4207.

Morel, F.M.M. and J.G. Hering, 1993. *Principles and Applications of Aquatic Chemistry*, a Wiley Interscience Publication, John Wiley and Sons, New York, 588 pages.

Morgan, I.J., R.P. Henry, and C.M. Wood, 1997. "The Mechanism of Acute Silver Nitrate Toxicity in Freshwater Rainbow Trout *(Oncorhynchus mykiss*) is Inhibition of Gill Na^+ and Cl^- Transport," *Aquatic Toxicology*, 38: 145-163.

Morgan, T.P., M. Grosell, K.C. Gilmour, R.C. Playle, and C.M. Wood, 2004. "Time Course Analysis of the Mechanism by which Silver Inhibits Active Na^+ and Cl^- Uptake in Gills of Rainbow Trout," *American Journal of Physiol. Regul. Integr. Comp. Physiol.*, 287: R234-R242.

Mount, D.R., D.D. Gulley, J.R. Hockett, T.D. Garrison, and J.M. Evans, 1997. "Statistical Models to Predict the Toxicity of Major Ions to *Ceriodaphnia dubia*, *Daphnia magna* and *Pimephales promelas* (Fathead Minnows)," *Environmental Toxicology and Chemistry*, 16(10): 2009-2019.

Naddy, R.B., J.W. Gorsuch, A.B. Rehner, G.R. McNerney, R.A. Bell, and J.R. Kramer, 2007. "Chronic Toxicity of Silver Nitrate to *Ceriodaphnia dubia* and *Daphnia magna*, and Potential Mitigating Factors," *Aquatic Toxicology*, 84: 1-10.

Naddy, R.B., G.R. McNerney, W.A. Stubblefield, R.A. Bell, R. Mathew, K.B. Wu, R. Santore, and P.R. Paquin, 2008. "Influence of Varying Water Quality Parameters on the Acute Toxicity of Silver to the Freshwater Cladoceran, *Ceriodaphnia dubia*," in preparation.

Pagenkopf, G.K., 1983. "Gill Surface Interaction Model for Trace-metal Toxicity to Fishes: Role of Complexation, pH, and Water Hardness," *Environmental Science and Technology*, 17(6): 342-347.

Paquin, P.R., D.M. Di Toro, R.C. Santore, D. Trivedi, and K.B. Wu, April 1999. "A Biotic Ligand Model of the Acute Toxicity of Metals. III. Application to Fish and *Daphnia* Exposure to Silver," Section 3 in *Integrated Approach to Assessing the Bioavailability and Toxicity of Metals in Surface Waters and Sediments*, a submission to the EPA Science Advisory Board, Office of Water, Office of Research and Development, Washington, DC, pp. 3-59 to 3-102. U.S. EPA - 822-E-99-001.

Paquin, P.R., J.W. Gorsuch, S. Apte, G.E. Batley, K.C. Bowles, P.G.C. Campbell, C.G. Delos, D.M. Di Toro, R.L. Dwyer, F. Galvez, R.W. Gensemer, G.G. Goss, C. Hogstrand, C.R. Janssen, J.C. McGeer, R.B. Naddy, R.C. Playle, R.C. Santore, U. Schneider, W.A. Stubblefield, C.M. Wood, and K.B. Wu, 2002a. "The Biotic Ligand Model: A Historical Overview," Special Issue: The Biotic Ligand Model for Metals – Current Research, Future Directions, Regulatory Implications, *Comparative Biochemistry and Physiology*, Part C 133(1-2): 3-35.

Paquin, P.R., V. Zoltay, R.P. Winfield, K.B. Wu, R. Mathew, R.C. Santore, and D.M. Di Toro, 2002b. "Extension of the Biotic Ligand Model of Acute Toxicity to a Physiologically-based Model of the Survival Time of Rainbow Trout (*Oncorhynchus mykiss)* Exposed to Silver," Special Issue: The Biotic Ligand Model for Metals–Current Research, Future Directions, Regulatory Implications, *Comparative Biochemistry and Physiology*, Part C 133(1-2): 305-343.

Paquin, P.R., J. Gorsuch, K.B. Wu, R. Mathew, R. Santore, and D.M. Di Toro, 2007. "Overview of the Silver Research and Development Program (1991-2006)", prepared for the International Imaging Industry Association (I3A), 131 pp.

Playle, R.C., R.W. Gensemer, and D.G. Dixon, 1992. "Copper Accumulation on Gills of Fathead Minnows: Influence of Water Hardness, Complexation and pH of the Gill Micro-Environment," *Environmental Toxicology and Chemistry*, 11(3): 381-391.

Playle, R.C., D.G. Dixon, and K. Burnison, 1993a. "Copper and Cadmium Binding to Fish Gills: Modification by Dissolved Organic Carbon and Synthetic Ligands," *Can. J. Fish. Aquat. Sci.*, 50: 2667-2677.

Playle, R.C., D.G. Dixon, and K. Burnison, 1993b. "Copper and Cadmium Binding to Fish Gills: Estimates of Metal-Gill Stability Constants and Modeling of Metal Accumulation," *Can. J. Fish. Aquat. Sci.*, 50: 2678-2687.

Playle, R.C., 1998. "Modelling Metal Interactions at Fish Gills," in *Paradigms of Trace Metal Bioaccumulation in Aquatic Ecosystems*, C.J. Watras, R.G. Carlton and D.B. Porcella, eds., a special issue of *Science of the Total Environment*, 219(2/3): 147-163.

Rickard, D.T. and G.W. Luther III, 2006. "Metal Sulfide Complexes and Clusters," *Reviews in Mineralogy & Geochemistry*, 61: 421-504.

Rozan, T.F., M.E. Lassman, D.P. Ridge, and G.W. Luther, III, 2000. "Evidence for Iron, Copper and Zinc Complexation as Multinuclear Sulphide Clusters in Oxic Rivers," *Nature*, 406: 879-882.

Rozan, T.F. and G.W. Luther, III, 2002. "Voltammetric Evidence Suggesting Ag Speciation is Dominated by Sulfide Complexation in River Water," *ACS Symposium Series 811, Environmental Electrochemistry, Analyses of Trace Element Biogeochemistry*, Published by the American Chemical Society.

Santore, R.C. and C.T. Driscoll, 1995. "The CHESS Model for Calculating Chemical Equilibria in Soils and Solutions," *Chemical Equilibrium and Reaction Models*, SSSA Special Publication 42, The Soil Society of America, American Society of Agronomy.

Santore, R.C., D.M. Di Toro, and P.R. Paquin, H.E. Allen and J.S. Meyer, 2001. "A Biotic Ligand Model of the Acute Toxicity of Metals. II. Application to Acute Copper Toxicity in Freshwater Fish and Daphnia," *Environmental Toxicology and Chemistry,* 20(10): 2397-2402.

Santore, R.C., R. Mathew, P.R. Paquin, and D.M. Di Toro, 2002. "Development of a Biotic Ligand Model of Acute Toxicity for Zinc," Special Issue: The Biotic Ligand Model for Metals–Current Research, Future Directions, Regulatory Implications, *Comparative Biochemistry and Physiology*, Part C 133(1-2): 271-285.

Santore, R., R. Bell, R. Mathew, P.R. Paquin, D.S. Smith, K. Wu and D.M. Di Toro, 2008. "Development of an Acute Biotic Ligand Model for Ag: Ag Speciation in Natural Waters with Organic Matter Interactions," in preparation.

Sikora, F.J. and F.J. Stevenson (1988). "Silver Complexation by Humic Substances: Conditional Stability Constants and Nature of Reactive Sites," *Geoderma*, 42: 353-363.

Skou, J.C., 1957. "The Influence of Some Cations on an Adenosine Triphosphatase from Peripheral Nerve," *Biochimica et Biophysica Acta*, 23: 394-401.

Smith, R.M. and A.E. Martell, 2001. "NIST Critically Selected Stability Constants of Metal Complexes, Version 6.0," U.S. Department of Commerce, Gaithersburg, MD 20899, USA.

Smith, D.S., R.A. Bell, and J.R. Kramer, 2002. "Metal Speciation in Natural Waters with Emphasis on Reduced Sulfur Groups as Strong Metal Binding Sites," Special Issue: The Biotic Ligand Model for Metals–Current Research, Future Directions, Regulatory Implications, *Comparative Biochemistry and Physiology*, Part C 133(1-2): 65-74.

Smith, D.S., R.A. Bell, J. Valliant, and J.R. Kramer, 2004. "Determination of Strong Ligand Sites in Sewage Effluent-Impacted Waters by Competitive Ligand Titration with Silver," *Environmental Science and Technology*, 38(7): 2120-2125.

Stephan, C.E., D.I. Mount, D.J. Hansen, J.H. Gentile, G.A. Chapman, and W.A. Brungs, January 1985. "Guidelines for Deriving Numerical National Water Quality Criteria for the Protection of Aquatic Organisms and Their Uses," U.S. EPA Office of Research and Development, Environmental Research Laboratories: Duluth, Minnesota; Narragansett, Rhode Island and Corvallis, Oregon, 98 pp.

Tipping, E., 1994. "WHAM - A Chemical Equilibrium Model and Computer Code for Waters, Sediments and Soils Incorporating a Discrete Site / Electrostatic Model of Ion Binding by Humic Substances," *Computers & Geosciences*, 20(6): 973-1023.

U.S. EPA, September 24, 1987. Draft "Ambient Aquatic Life Water Quality Criteria for Silver," Office of Research and Development, Environmental Research Laboratories, Narragansett, Rhode Island.

U.S. EPA, April 6-7, 1999a. *Integrated Approach to Assessing the Bioavailability and Toxicity of Metals in Surface Waters and Sediments*, a report to the EPA Science Advisory Board, Office of Water, Office of Research and Development, Washington, D.C., EPA-822-E-99-001.

U.S. EPA, October 28, 1999b. "Notice of Intent to Revise Aquatic Life Criteria for Copper, Silver, Lead, Cadmium, Iron and Selenium," *Federal Register*, Volume 64, Number 209, Document 99-28314, pp. 58409-58410.

U.S. EPA, February 2000. "An SAB Report: Review of the Biotic Ligand Model of the Acute Toxicity of Metals," prepared by the Ecological Processes and Effects Committee of the Science Advisory Board, EPA-SAB-EPEC-00-006.

U.S. EPA, October 2002a. "Methods for Measuring the Acute Toxicity of Effluents and Receiving Waters to Freshwater and Marine Organisms," Fifth Edition, EPA-821-R-02-012.

U.S. EPA, 2002b. Draft "The Biotic Ligand Model: Technical Support Document for Its Application to the Evaluation of Water Quality Criteria for Copper," Water Quality Standards Branch, Office of Science and Technology, U.S. Environmental Protection Agency, Washington, D.C., EPA, February 22, 2002.

U.S. EPA, 2007. "2007 Update of Ambient Water Quality Criteria for Copper," United States Environmental Protection Agency, Office of Water, Office of Science and Technology, EPA 822-R-03-026.

Van Genderen, E.J., A.C. Ryan, J.R. Tomasso, and S.J. Klaine, 2003. "Influence of Dissolved Organic Matter Source on Silver Toxicity to *Pimephales promelas*," *Environmental Toxicology and Chemistry*, 22(11): 2746-2751.

Van Genderen, E.J., A.C. Ryan, J.R. Tomasso, and S.J. Klaine, 2005. "Evaluation of Acute Copper Toxicity to Larval Fathead Minnows (*Pimephales promelas*) in Soft Surface Waters," *Environmental Toxicology and Chemistry*, 24(2): 408-414.

Webb, N.A. and Wood, C.M., 1998. "Physiological Analysis of the Stress Response Associated with Acute Silver Nitrate Exposure in Freshwater Rainbow Trout (*Oncorhynchus mykiss*)," *Environmental Toxicology and Chemistry*, 17: 579–588.

Wilcox, S.J. and T.H. Dietz, 1995. "Potassium Transport in the Freshwater Bivalve *Dreissena polymorpha*," *The Journal of Experimental Biology*," 198: 861-868.

Wood, C.M., C. Hogstrand, F. Galvez, and R.S. Munger, 1996. "The Physiology of Waterborne Silver Toxicity in Freshwater Rainbow Trout (*Oncorhynchus mykiss*): 1. The Effects of Ionic Ag^{+}," *Aquatic Toxicology*, 35: 93-109.

Wu, K.-B., R. Mathew, G. Bielmyer, K. Brix, R. Naddy, R. Bell, M.Grosell, P.R. Paquin, R. Santore, and D.M. Di Toro, 2008. "Development of an Acute Biotic Ligand Model for Silver: Application to Fish and Invertebrates," in preparation.

WASTEWATER UTILITY

Alabama
Montgomery Water Works & Sanitary Sewer Board

Alaska
Anchorage Water & Wastewater Utility

Arizona
Glendale, City of, Utilities Department
Mesa, City of
Peoria, City of
Phoenix Water Services Dept.
Pima County Wastewater Management
Safford, City of
Tempe, City of

Arkansas
Little Rock Wastewater Utility

California
Central Contra Costa Sanitary District
Corona, City of
Crestline Sanitation District
Delta Diablo Sanitation District
Dublin San Ramon Services District
East Bay Dischargers Authority
East Bay Municipal Utility District
Eastern Municipal Water District
El Dorado Irrigation District
Fairfield-Suisun Sewer District
Fresno Department of Public Utilities
Inland Empire Utilities Agency
Irvine Ranch Water District
Las Virgenes Municipal Water District
Livermore, City of
Los Angeles, City of
Los Angeles County, Sanitation Districts of
Napa Sanitation District
Orange County Sanitation District
Palo Alto, City of
Riverside, City of
Sacramento Regional County Sanitation District
San Diego Metropolitan Wastewater Department, City of
San Francisco, City & County of
San Jose, City of
Santa Barbara, City of
Santa Cruz, City of
Santa Rosa, City of
South Bayside System Authority
South Coast Water District
South Orange County Wastewater Authority
Stege Sanitary District
Sunnyvale, City of
Union Sanitary District
West Valley Sanitation District

Colorado
Aurora, City of
Boulder, City of
Greeley, City of
Littleton/Englewood Water Pollution Control Plant
Metro Wastewater Reclamation District, Denver

Connecticut
Greater New Haven WPCA
Stamford, City of

District of Columbia
District of Columbia Water & Sewer Authority

Florida
Broward, County of
Fort Lauderdale, City of
Jacksonville Electric Authority (JEA)
Miami-Dade Water & Sewer Authority
Orange County Utilities Department
Reedy Creek Improvement District
Seminole County Environmental Services
St. Petersburg, City of
Tallahassee, City of
Tampa, City of
Toho Water Authority
West Palm Beach, City of

Georgia
Atlanta Department of Watershed Management
Augusta, City of
Clayton County Water Authority
Cobb County Water System
Columbus Water Works
Fulton County
Gwinnett County Department of Public Utilities
Savannah, City of

Hawaii
Honolulu, City & County of

Idaho
Boise, City of

Illinois
Decatur, Sanitary District of
Greater Peoria Sanitary District
Kankakee River Metropolitan Agency
Metropolitan Water Reclamation District of Greater Chicago
Wheaton Sanitary District

Iowa
Ames, City of
Cedar Rapids Wastewater Facility
Des Moines, City of
Iowa City

Kansas
Johnson County Wastewater
Unified Government of Wyandotte County/ Kansas City, City of

Kentucky
Louisville & Jefferson County Metropolitan Sewer District
Sanitation District No. 1

Louisiana
Sewerage & Water Board of New Orleans

Maine
Bangor, City of
Portland Water District

Maryland
Anne Arundel County Bureau of Utility Operations
Howard County Bureau of Utilities
Washington Suburban Sanitary Commission

Massachusetts
Boston Water & Sewer Commission
Massachusetts Water Resources Authority (MWRA)
Upper Blackstone Water Pollution Abatement District

Michigan
Ann Arbor, City of
Detroit, City of
Holland Board of Public Works
Saginaw, City of
Wayne County Department of Environment
Wyoming, City of

Minnesota
Rochester, City of
Western Lake Superior Sanitary District

Missouri
Independence, City of
Kansas City Missouri Water Services Department
Little Blue Valley Sewer District
Metropolitan St. Louis Sewer District

Nebraska
Lincoln Public Works and Utilities Department

Nevada
Henderson, City of
Las Vegas, City of
Reno, City of

New Jersey
Bergen County Utilities Authority
Ocean County Utilities Authority
Passaic Valley Sewerage Commissioners

New York
New York City Department of Environmental Protection

North Carolina
Charlotte/Mecklenburg Utilities
Durham, City of
Metropolitan Sewerage District of Buncombe County
Orange Water & Sewer Authority
University of North Carolina, Chapel Hill

Ohio
Akron, City of
Butler County Department of Environmental Services
Columbus, City of
Metropolitan Sewer District of Greater Cincinnati
Montgomery, County of
Northeast Ohio Regional Sewer District
Summit, County of

Oklahoma
Oklahoma City Water & Wastewater Utility Department
Tulsa, City of

Oregon
Albany, City of
Clean Water Services
Eugene, City of
Gresham, City of
Portland, City of Bureau of Environmental Services
Water Environment Services

Pennsylvania
Hemlock Municipal Sewer Cooperative (HMSC)
Philadelphia, City of
University Area Joint Authority

South Carolina
Charleston Water System
Mount Pleasant Waterworks & Sewer Commission
Spartanburg Water

Tennessee
Cleveland Utilities
Knoxville Utilities Board
Murfreesboro Water & Sewer Department
Nashville Metro Water Services

Texas
Austin, City of
Dallas Water Utilities
Denton, City of
El Paso Water Utilities
Fort Worth, City of
Houston, City of
San Antonio Water System

WERF SUBSCRIBERS

Trinity River Authority

Utah

Salt Lake City Corporation

Virginia

Alexandria Sanitation Authority
Arlington, County of
Fairfax County
Hampton Roads Sanitation District
Hanover, County of
Henrico, County of
Hopewell Regional Wastewater Treatment Facility
Loudoun Water
Lynchburg Regional WWTP
Prince William County Service Authority
Richmond, City of
Rivanna Water & Sewer Authority

Washington

Everett, City of
King County Department of Natural Resources
Seattle Public Utilities
Sunnyside, Port of
Yakima, City of

Wisconsin

Green Bay Metro Sewerage District
Kenosha Water Utility
Madison Metropolitan Sewerage District
Milwaukee Metropolitan Sewerage District
Racine, City of
Sheboygan Regional Wastewater Treatment
Wausau Water Works

Australia

ACTEW (Ecowise)
South Australian Water Corporation
South East Water Limited
Sydney Water Corporation
Water Corporation of Western Australia

Canada

Edmonton, City of/Edmonton Waste Management Centre of Excellence
Lethbridge, City of
Regina, City of, Saskatchewan
Toronto, City of, Ontario
Winnipeg, City of, Manitoba

New Zealand

Watercare Services Limited

STORMWATER UTILITY

California

Fresno Metropolitan Flood Control District
Los Angeles, City of, Department of Public Works
Monterey, City of
San Francisco, City & County of
Santa Rosa, City of
Sunnyvale, City of

Colorado

Aurora, City of
Boulder, City of

Florida

Orlando, City of

Georgia

Griffin, City of

Iowa

Cedar Rapids Wastewater Facility
Des Moines, City of

Kansas

Overland Park, City of

Kentucky

Louisville & Jefferson County Metropolitan Sewer District

Maine

Portland Water District

North Carolina

Charlotte, City of, Stormwater Services

Pennsylvania

Philadelphia, City of

Tennessee

Chattanooga Stormwater Management

Texas

Harris County Flood Control District, Texas

Washington

Bellevue Utilities Department
Seattle Public Utilities

STATE

Connecticut Department of Environmental Protection
Kansas Department of Health & Environment
New England Interstate Water Pollution Control Commission (NEIWPCC)
Ohio River Valley Sanitation Commission
Urban Drainage & Flood Control District, CO

CORPORATE

ADS Environmental Services
Advanced Data Mining International
Alan Plummer & Associates
Alpine Technology Inc.
Aqua-Aerobic Systems Inc.
Aquateam–Norwegian Water Technology Centre A/S
ARCADIS
Associated Engineering
Black & Veatch
Blue Water Technologies, Inc.
Boyle Engineering Corporation
Brown & Caldwell
Burgess & Niple, Ltd.
Burns & McDonnell
CABE Associates Inc.
The Cadmus Group
Camp Dresser & McKee Inc.
Carollo Engineers Inc.
Carpenter Environmental Associates Inc.
Carter & Burgess, Inc.
CET Engineering Services
CH2M HILL
CRA Infrastructure & Engineering
CONTECH Stormwater Solutions
D&B/Guarino Engineers, LLC
Damon S. Williams Associates, LLC
Earth Tech Inc.
Ecovation
EMA Inc.
Environmental Operating Solutions, Inc.
Environ/The ADVENT Group, Inc.
Fay, Spofford, & Thorndike Inc.
Freese & Nichols, Inc.
ftn Associates Inc.
Gannett Fleming Inc.
Garden & Associates, Ltd.
Geosyntec Consultants
GHD
Golder Associates Ltd.
Greeley and Hansen LLC
Hazen & Sawyer, P.C.
HDR Engineering Inc.
HNTB Corporation
Hydromantis Inc.
HydroQual Inc.
Infilco Degremont Inc.
Jacques Whitford NAWE, Inc.
Jason Consultants LLC Inc.
Jordan, Jones, & Goulding Inc.
KCI Technologies Inc.
Kelly & Weaver, P.C.
Kennedy/Jenks Consultants
KMK Consultants
Larry Walker Associates
Limno-Tech Inc.
The Low Impact Development Center Inc.
Malcolm Pirnie Inc.
Material Matters
McKim & Creed
Metcalf & Eddy Inc.
MPR Engineering Corporation, Inc.
MWH
O'Brien & Gere Engineers Inc.
Odor & Corrosion Technology Consultants Inc.
Paradigm Environmental Technologies, Inc.
Parametrix Inc.
Parsons
Post, Buckley, Schuh & Jernigan
Praxair, Inc.
Ring Industrial Group
RMC Water & Environment
Ross & Associates Ltd.
SAIC
Siemens Water Technologies
The Soap & Detergent Association
Smith & Loveless, Inc.
Stantec Consulting Inc.
Stearns & Wheler, LLC
Stone Environmental Inc.
Stratus Consulting Inc.
Synagro Technologies Inc.
Tetra Tech Inc.
Trojan Technologies Inc.
Trussell Technologies, Inc.
Uni-Bell PVC Pipe Association
URS Corporation
Westin Engineering Inc.
Weston Solutions Inc.
Woodard & Curran
Zoeller Pump Company

INDUSTRY

American Electric Power
American Water
Anglian Water Services, Ltd.
ChevronTexaco Energy Research & Technology Company
The Coca-Cola Company
Dow Chemical Company
DuPont Company
Eastman Chemical Company
Eli Lilly & Company
Merck & Company Inc.
Procter & Gamble Company
Suez Environnment
United Utilities North West (UUNW)
United Water Services LLC

Note: List as of 8/11/08

IWA Publishing's authorised EU representative for General Product Safety Regulations is Diane D'Arras, 15 rue Duret, 75116 Paris, France, e-mail: safety@iwap.co.uk.

Printed and bound by CPI Group (UK) Ltd, Croydon, CR0 4YY
12/05/2026
02108881-0010